Army Medical Support for Peace Operations and Humanitarian Assistance

Lois M. Davis

Susan D. Hosek

Michael G. Tate

Mark Perry

Gerard Hepler

Paul S. Steinberg

Prepared for the
United States Army

Arroyo Center

RAND

This report examines issues confronting the U.S. Army and the Army Medical Department (AMEDD) in providing medical support for "operations other than war" (OOTW)—a broad range of missions including peacekeeping, peace enforcement, humanitarian assistance, disaster relief, and nation assistance, among others. In this study we deal specifically with providing medical support for peace and humanitarian operations. Such operations often impose heavy demands on the United States for medical personnel, equipment and supplies, patient evacuation, and other scarce health care resources. It is important, therefore, for the Army to understand the nature of these demands and strategies for dealing with them, given the increasing frequency of OOTW since the end of the Cold War.

This report reviews U.S. military medical experience with several recent OOTW, focusing primarily on the UNPROFOR mission in the Balkans and the operation in Somalia. It identifies the special features of the medical support in such operations (as distinct from support of combat operations) and issues unique to supporting a multinational force as part of a coalition. It also suggests steps that the Army, the other military services, and the U.S. government could take to improve our ability to respond to such operations, to limit the demands that they may impose, and to minimize their impact on the Army's readiness mission and on peacetime health care delivery. The findings of this study will be of interest not only to the Army and the AMEDD, but also to the other military Surgeons General and medical departments.

This research was sponsored by The U.S. Army Surgeon General, as part of a project on future Army medical structure. It was conducted in the Manpower and Training Program of RAND's Arroyo Center, a federally funded research and development center sponsored by the United States Army.

CONTENTS

Appendix

As the United States contends with the strategic uncertainty in the post–Cold War era, it must consider the role of its military forces in operations other than war (OOTW), such as peacekeeping, peace enforcement, or humanitarian assistance. In these operations, medical issues tend to play a more central role than in combat operations and the medical support requirements tend to be broader, particularly if a multinational force is involved. Hence, OOTW may place greater and new demands on the Army Medical Department (AMEDD).

This report synthesizes recent military experience in medical support for OOTW, and it recommends ways for the Army to improve OOTW support while minimizing the impact of OOTW on the AMEDD's wartime readiness mission and on peacetime health care delivery to its beneficiaries. It deals primarily with five issues:

- How do the medical support requirements of OOTW differ from those of combat missions? What special demands are imposed by OOTW, especially when multinational forces are involved?

- How can the Army manage the inherent pressures toward open-ended expansion of the medical mission in OOTW?

- How can the AMEDD build a robust and flexible system to meet the broad range of demands associated with these operations?

- How can the Army minimize the impact of OOTW on the Army's readiness mission and its ability to provide beneficiary care?

- What kind of planning, education, and training may be required to better prepare the Army to support OOTW in the future?

To address these issues, we examined two main deployments: the UN operation in the Balkans in 1992–1994 and the U.S.-led mission to Somalia in 1993–1994. We also conducted a less extensive analysis of the U.S.-led mission to Haiti in 1994–1995. For each deployment, our main sources were documents and interviews with key participants and officials, supplemented by quantitative data on unit configurations and patient- and provider-level data on deploying medical units.

NATURE OF OOTW MEDICAL DEMAND

The demand for medical services in OOTW differs in some important ways from that in combat operations. First, the patient population tends to be much broader, with more diverse treatment needs. In addition to U.S. troops, Army medical units may be called upon to treat (a) local civilians, (b) refugees, (c) troops of coalition partners, and (d) employees of the U.S. government, UN, NATO, or civilian contractors. These patient groups vary more than U.S. troops in their health status, age structure, proportion of females, and type of acute or chronic medical conditions requiring treatment in-theater. In addition, troops in a multinational coalition force tend to have lower levels of predeployment medical screening, preventive medicine support, and medical and dental readiness—increasing medical support requirements. These differences also mean that the AMEDD may be called upon to provide a broader range of services (including pediatric and ob/gyn care) in these operations and must be prepared to treat certain infectious diseases and chronic medical conditions not common among U.S. forces. Thus, in OOTW the demand for medical services is often closer to what a community hospital would face, compared to a military hospital in support of combat operations, which is geared primarily toward trauma and emergency care.

Second, other available medical assets may be deficient. For instance, the existing medical infrastructure of the host nation may have been destroyed. Coalition partners' own medical assets may be inadequate for the mission; troops in a multinational force will differ in the level and type of equipment, supplies, and training of their medical personnel and units, as well as in the quality of care these assets provide. As a result, the United States may feel compelled to compensate for these differences by plugging holes in the theater medical system, supplementing other troops' medical assets, and imposing U.S. standards of care.

Third, some coalition patients may require intensive or prolonged hospital care that goes beyond what military health service support is designed to provide in a theater of operations. To begin with, some patients may be in poor condition owing to lack of preventive medicine in the field, inadequate Echelons I or II care by their organic medical assets, or delays in transport to the U.S. military hospital. For example, in the Balkans the forward surgical teams within the UN force provided widely varying quality of care and uneven coverage across different sectors. The U.S. hospital may also find it difficult to repatriate some coalition patients if adequate treatment is not available in their home countries. Indeed, some countries' deficiencies in quality of care may serve as an incentive for them to leave their soldiers at a U.S. military hospital.

Fourth, many OOTW missions have a humanitarian component, including public health actions or prevention (e.g., ensuring the quality of the local water supply to prevent the spread of cholera, or establishing basic sanitation conditions within a refugee camp or community hospital). In such situations, the Army may find itself providing medical supplies, community health services, public health education, training, and even basic equipment to shore up the local medical infrastructure.

Fifth, although OOTW require a broader range of services, patient demand tends to be relatively low. The size of the hospital required is often small (averaging 60 beds and 120–140 medical staff in recent OOTW). This suggests that the military hospital may easily be overwhelmed in a mass-casualty situation and that medical evacuation will become a top priority. For instance, Somalia was an example of the AMEDD doing its combat mission in an OOTW environment; patient demand was relatively low, characterized by peaks and valleys yet always with the potential for combat. That theater illustrated not only the low-end requirements in OOTW, but also the difficulty of planning medical support for these types of missions. A key lesson from Somalia may have been to staff for a little more than the average and then ensure the capability to extend for the surge.

Overall, we found the critical determinants of the medical support requirements in OOTW to include (a) the presence of refugee or displaced populations, (b) whether there is a humanitarian component to the operation, (c) the degree to which the host nation's medical infrastructure has been compromised, (d) whether the United States is acting unilaterally or with a multinational force, (e) the level of support the United States has been tasked to provide to a multinational force, (f) differences in medical readiness among coalition troops, and (g) the degree of variability in coalition partners' medical assets. Because of these features, OOTW can present a broad range of resource demands with rapidly changing mission requirements, suggesting the need for flexibility in planning and ability to tailor support to the mission.

MISSION EXPANSION

Such conditions generate both internal and external pressures for the medical mission to expand—a phenomenon often called "mission creep." In OOTW the demand for services is often open-ended and has the potential to consume large amounts of medical resources, undermining Army readiness for other missions. Some key factors that push toward a larger mission include the following:

Needs of coalition partners. A broader set of treatment demands arise among soldiers from other nations. Some coalition troops may utilize the theater medical system in ways it was not intended, and coalition partners' own medical assets may be substandard.

Demand induced by U.S. actions. The U.S. informal policy of "if we hurt them, we fix them" leads to involvement with civilian populations in any event. In addition, U.S. soldiers may bring in sick or injured civilians to the military hospital for care.

Excess capacity. In OOTW, excess medical capacity is unavoidable to a large extent, given the wide fluctuations in patient demand and the fluid nature of these operations. As a result, the in-theater medical facility may be underutilized at times. This available, but unused, supply tends to stimulate demand. In addition, military medical units often find themselves in areas with overwhelming medical need among the local civilian population. Providers also want to continue practicing their specialties to maintain their clinical skills.

Outside requests and influences. The UN, coalition partners, foreign ambassadors, other U.S. agencies, and the State Department often urge the Army to expand the medical mission. Coalition partners also may define a broader medical mission and set of objectives for themselves, creating a disparity that pushes the United States in a similar direction or sets up unrealistic expectations of the U.S. military's medical role.

Ethical and professional considerations. Medical personnel have a professional orientation that implies an obligation to help with urgent medical problems (among civilian and coalition populations) and an understandable desire to respond to medical need, regardless of the formal mission parameters.

Inadequately defined missions. Guidance from the strategic or interagency levels may not adequately specify who is entitled to what type of care in OOTW. The lack of articulation of a national medical strategy for OOTW that defines the objectives and medical rules of engagement has led to ambiguity, which in turn encourages mission creep.

Given the above factors, this report describes several actions the Army and other U.S. agencies might take to bound the medical mission appropriately. These include:

- Clarifying up front the medical mission, its objectives, desired end state, and classes of patients eligible for services;

- Limiting treatment of civilians to the level of care customary in the region and not imposing U.S. standards of care that the host nation or nongovernmental organizations (NGOs) are unable to sustain once the U.S. military departs;

- Addressing repatriation problems by establishing procedures for evacuating coalition patients to their home countries, facilities in neighboring countries, or local hospitals;

- Relying on civilian contractors or negotiating workload with local medical facilities and NGOs in areas where they can help or have a comparative advantage over the U.S. military.

A ROBUST AND FLEXIBLE STRUCTURE

Because OOTW tend to encompass a broad range of medical tasks but require less total capacity than combat missions, it does not make sense for the AMEDD to construct new structure for these operations. Rather, the key is to build a robust and flexible structure that will allow the AMEDD to respond to the breadth of demands it now faces. The report details several recommendations for structural improvements, including the following:

Modular units, consistent with the concept of Medical Force XXI, and tailorable for the demands of OOTW.

A deployable isolation ward capability for treating patients with serious contagious diseases such as tuberculosis.

Extended preventive medicine and physical therapy services, as well as limited rehabilitative capabilities to treat land mine injuries and sports-related injuries (both common in OOTW).

Development of staging teams or other means of ensuring a surge capability to handle and quickly evacuate casualties in a mass-casualty situation that might overwhelm a small hospital.

Special-purpose support packages for geriatric, gynecological, and pediatric care. Also, minimal-care wards for housing soldiers or translators who accompany injured or sick coalition soldiers, adults who accompany a child, orphans, and coalition patients who may no longer require care but for whom there are delays in repatriating them back to their unit or home country.

Extended use of telemedicine capabilities has the potential to play an important role in OOTW, where in-theater personnel ceilings may prevent deploying the many different specialties required to treat the full range of diseases and medical conditions encountered. Innovative uses of this technology also need to be explored. For example, telemedicine may be able to play a role in addressing repatriation problems and meeting the expanded medical intelligence and linguistic requirements associated with OOTW.

MINIMIZING IMPACT ON READINESS AND PEACETIME CARE

Several factors give OOTW the potential to affect future wartime readiness and peacetime health care delivery: the simultaneity of demands, their open-ended character, and the Army's direction that it support them without any degradation in beneficiary care. To preserve its medical support capabilities in the face of OOTW demands, the Army may want to consider designating certain medical units as OOTW hospitals and staffing those hospitals with two of each of the most critical functional elements. For example, of the 13 CONUS TOE hospitals currently in the active-duty structure, the AMEDD could build one or several into a "1.5" hospital. Then, if half of a hospital deploys on an OOTW, a complete hospital would still be available for a second deployment.

Such units would then know in advance that they will be on the "hot seat" for supporting OOTW, and this designation could be rotated among active-duty hospitals on a yearly basis. This would allow the Army to avoid pulling personnel from a number of different military treatment facilities to support a single deployment, and it would open up the possibility for advance planning to maintain beneficiary care while supporting an OOTW. For example, military treatment facilities might rely on standing contracts with civilian providers or place deployable PROFIS personnel in noncritical positions to minimize the impact of these deployments on peacetime health care.

PLANNING, EDUCATION, AND TRAINING

Flexible planning. First and most critically, Army planners need to consider not only the medical requirements for supporting a deploying force, but also the broader missions the Army may be assigned in OOTW. For example, in operations involving humanitarian assistance or refugee populations, the medical mission is likely to be broader than the basic workload of supporting the deploying force. In fact, the real thrust of the AMEDD's workload may be to provide health support to the host nation. Yet planners continue to view the medical mission as limited to its traditional combat service support role. This in turn has led to a mismatch between the requirement and the force provided, and at times to the inefficient use of medical assets. To avoid such problems, we offer three recommendations.

First, planning should recognize the special medical requirements related to civilian populations and multinational forces. For example, lack of preventive medicine and poor quality of care by some coalition troops' medical teams may mean that a U.S. Echelon III hospital receives more patients whose treatment is complicated or resource-intensive. And a political reality of OOTW is that the U.S. military may be tasked to provide care to civilians—not only local people, but refugees, UN and NATO employees, and coalition soldiers—whether or not it is part of the official medical mission.

Second, planning should explicitly consider the varied types of at-risk populations, categories of patients, and medical conditions enumerated above. Although it is not possible to plan for all eventualities, in general we can do better at predicting the population at risk and the key determinants of the support requirements in OOTW.

Third, planning should include advance assessment teams with OOTW expertise. Such teams should include not only physicians with prior OOTW deployment experience, but also preventive medicine officers, community health nurses, and others with special expertise for OOTW.

Individual education and training. Clearly there is a large political element associated with OOTW. Yet many of the AMEDD officers and enlisted personnel who have deployed on these missions have had little experience in political matters. This encourages mission creep at the individual and tactical level. Such tendencies could be combated by better education and training.

For AMEDD officers, the Officer Basic and Advanced Courses could include an introduction to OOTW with a focus on problem-solving exercises, and the Command and General Staff College could provide a forum for discussions on medical support requirements, public health issues, and other problems medical units face in OOTW. The Army War College curriculum could include coursework on planning and leading OOTW, as well as on policy and political issues. In addition, the AMEDD may want to articulate humanitarian rules of engagement for enlisted and nonmedical officers in OOTW.

Integrated medical and line training exercises. To educate both line and medical officers, Army medical units could become more involved in collective training for OOTW at the JRTC. It is in such a training environment that line and Army medical officers could hash out many of the medical decisions and practices associated with OOTW prior to a deployment, rather than relying on ad hoc decisionmaking in the theater. Further, medical commanders and their core staff could receive training on interpreting an operation plan, developing a tactical plan, and making the kind of clinical and command decisions they may face in a Somalia, Bosnia, or Haiti scenario. Since medical units in the past have had limited participation in JRTC training exercises (e.g., providing site support), it will be up to the AMEDD to articulate a future training strategy that will incorporate this type of experience.

Medical-unique training. Physicians, nurses, and other providers also need training in the types of treatment dilemmas they may encounter in OOTW, on how to respond appropriately to help avert the tendency toward assuming an additional mission, and on how actions undertaken at the delivery end may inadvertently lead to an expansion of the medical mission. In addition to JRTC training, one way to accomplish this would be for the AMEDD to undertake medical-unique training for OOTW at Camp Bullis to help medical personnel anticipate the demands and treatment decisions they will confront.

REQUIREMENTS OF COALITION OPERATIONS

Coalition operations pose unique problems in providing and structuring medical support. In general, the U.S. military tends to serve as the backbone of the medical support in OOTW, regardless of whether the mission is to support U.S. troops or a multinational force.

Instead of being able to set up an integrated theater medical system with consistent quality across echelons of care, the U.S. military often faces a structure of highly variable quality, with holes and gaps among the different elements. Given this, the United States and its other key coalition partners may want to take the lead in developing a revised definition of echelons of care, specific to OOTW involving a multinational force. Such a plan would set standards for medical readiness, unit readiness, training, equipment, and standards of care, as well as a realistic evacuation policy.

International differences in standards of care and medical practice raise serious questions for U.S. medical support. For example, how much quality variation can the United States afford in the theater medical system—and what are the attendant risks? Few militaries are as aggressive as that of the United States in such areas as trauma care, and some have substantially lower standards of care. Although one may be able to ensure quality in a clinic setting, U.S. military medical personnel may not be the first assets to reach a wounded U.S. soldier in an emergency situation during a coalition operation. This suggests that the United States may not be able to afford much variability in the theater medical assets. Instead, it may want to serve as the coordinator of medical care. Or the United States may want to continue to impose its standards on other coalition forces. If either option is selected, the United States needs to be explicit about it and negotiate compensation up front for these additional activities.

In general, the United States can expect to be tasked increasingly to provide air evacuation and medical logistics in multinational operations, since it has one of the few militaries with such capabilities. And, as illustrated by the experience of all three services in the Balkans and Haiti, the U.S. military must rely on its own logistics support rather than UN systems, because of quality problems and differences in standards.

OVERALL CONCLUSIONS

As the Army and the AMEDD, like the rest of the U.S. military, continue to downsize, no one can clearly envision the strategic environment they will be operating within in the future. We do anticipate, however, that the United States will continue to undertake OOTW, perhaps at an increasing rate.

In general, we found OOTW to entail a broader set of demands upon the medical component. Planning for future OOTW needs to recognize the breadth of such demands, especially in multinational operations. The AMEDD will need to ensure a broad-based flexibility to

support the diversity of new missions it will be called upon to undertake in an OOTW environment. At the same time, though, given the
overwhelming medical need and the fact that the U.S. military often
serves as the backbone of the medical support in OOTW, the United
States needs to focus and contain its medical involvement in these
missions where possible. Finally, many of the medical issues identified here are systemic—to be confronted successfully, they need to
be addressed not only at the Army headquarters level, but also at the
strategic, operational, and tactical levels.

ACKNOWLEDGMENTS

We would like to express our gratitude to Lieutenant General Alcide M. LaNoue, The Surgeon General of the U.S. Army, and Major General Thomas R. Tempel, Deputy Surgeon General of the U.S. Army, the sponsors of this study. We also are grateful to Major General James B. Peake and Brigadier General Harold L. Timboe for their valuable input to this report.

In addition, we would like to acknowledge the assistance of the staff officers of the Office of The Army Surgeon General for providing access to unpublished data and for sharing their own experiences in planning or undertaking the operations examined in this report. In particular, we would like to acknowledge the valuable input of COL John W. Zurcher, COL C. Greg Stevens, COL Robert J. Poux, COL Gerald A. Palmer, LTC David S. Heintz, LTC David W. Williams, MAJ Jeffrey M. Unger, MAJ Michael C. Gunn, and the staff officers within the Directorate of Health Care Operations, Office of The Surgeon General.

Our thanks are also due to COL Paul S. Beaty for shepherding this product through and for ensuring that we received timely input from various elements within the Army Medical Department (AMEDD). As a result of his efforts, this report has benefited from the careful review of a number of individuals within the AMEDD. Specifically, we would like to acknowledge the 44th Medical Brigade, COL Dale Carroll, COL Lester Martinez-Lopez, LTC Deogracia Quinones, and LTC John A. Haynie. In addition, we would like to express our gratitude to COL David Nolan for his valuable input and thoughtful review of an earlier draft of this report.

We interviewed a number of AMEDD personnel who had participated in these operations and who gave freely of their time, without
whose insights this report would not have been possible. Although
they are too numerous to mention here, we gratefully acknowledge
their input to this study. We also express our thanks to the individual
commanders interviewed for their in-depth comments and candid
assessments of the various operations and lessons learned. We are
grateful to the Navy and Air Force officers and commanders who
supported the U.S. military's medical mission in the Balkans and
whose valuable insights on this operation significantly contributed to
understanding the complex issues associated with coalition operations. Lastly, we thank the medical and dental personnel from
Womack Army Medical Center and the Fort Bragg Dental Activity.

Patient and provider data for the two case studies were obtained
from the Directorate of Patient Administrations Systems and Biostatistics Activities, AMEDD Center and School, Fort Sam Houston,
Texas. We thank COL Stuart Baker, Director, Mrs. Francie Mc-
Queeney, and Mrs. Deborah Ferrell for providing access to these data
and in a format needed for our analyses. We also acknowledge CAPT
James Carlisle, Naval Medical Center, San Diego, California, who
shared patient data from the Navy's Fleet Hospital 6.

This report has greatly benefited from the thoughtful review and
comments of our two reviewers, COL (ret.) Paul DeBree and RAND
colleague Jed Peters, which we gratefully acknowledge. Also, we
would like to thank our RAND colleagues Michael Polich and Ron
Sortor for their input and guidance throughout the course of this research.

AC	Active Component
AF	Air Force
AIDS	Acquired Immune Deficiency Syndrome
AIREVAC	Air Evacuation
AMEDD	Army Medical Department
ANCOC	Advanced Non-Commissioned Officer Course
AOCs	Areas of Concentration
ATH	Air Transportable Hospital
BNCOC	Basic Non-Commissioned Officer Course
CAS3	Combined Arms and Services Staff School
C4I	Command, Control, Communications, and Intelligence
CENTCOM	Central Command
CHAMPUS	Civilian Medical Program of the Uniformed Services
CINC	Commander in Chief
CJTF-S	Combined Joint Task Force Somalia
CMOC	Civil/Military Operations Center
CMS	Central Material Service

CONUS	Continental United States
CSA	Chief of Staff of the Army
CSH	Combat Support Hospital
DEPMEDS	Deployable Medical System
DNBI	Disease and Nonbattle Injury
DNR	Do-Not-Resuscitate
DoD	U.S. Department of Defense
ER	Emergency Room
EUCOM	European Command
EVAC	Evacuation Hospital
FDA	Federal Drug Administration
FH	Field Hospital
FM	Field Manual
FORSCOM	Forces Command
FST	Forward Surgical Teams
FWD	Forward
FYROM	Former Yugoslav Republic of Macedonia
GAO	U.S. Government Accounting Office
GI	Gastrointestinal
GME	Graduate Medical Education
HACC	Humanitarian Assistance Coordination Center
HIV	Human Immune Deficiency Virus
HROs	Humanitarian Relief Organizations
ICU	Intensive Care Unit
IMA	Individual Mobilization Augmentee
IOM	International Organization of Migration
JCS	Joint Chiefs of Staff

JRTC	Joint Readiness Training Center
JTF	Joint Task Force
JTF-PP	Joint Task Force Provide Promise
JTF-S	Joint Task Force-Somalia
JTFSC-S	Joint Task Force Support Command-Somalia
LOS	Length of Stay
MARFOR	Marine Forces
MASF	Mobile Aeromedical Staging Facility
MASH	Mobile Army Surgical Hospital
MEDCEN	Medical Center
MEDCOM	Medical Command
MEDDAC	Medical Activity
MEDEVAC	Medical Evacuation
MEDLOG	Medical Logistics
MFO Sinai	Multinational Force, Sinai
MNF	Multinational Force
MOS	Military Occupational Specialty
MP	Military Police
MSC	Medical Service Corps
MTF	Medical Task Force
MTFs	Military Treatment Facilities
NATO	North Atlantic Treaty Organization
NCOs	Noncommissioned Officers
NDMS	National Disaster Medical System
NGO	Nongovernmental Organization
NTC	National Training Center
OAC	Officers Advanced Course

OBC	Officers Basic Course
OB/GYN	Obstetrics and Gynecology
OCH	Operation Continue Hope
OCONUS	Outside CONUS
ODS	Operation Desert Storm
OFDA	Office of Federal Disaster Assistance, U.S. Department of State
OOTW	Operations Other Than War
OPP	Operation Provide Promise
OPR	Operation Provide Relief
OPTEMPO	Operating Tempo
ORH	Operation Restore Hope
OASD	Office of the Assistant Secretary of Defense
OT/PT	Occupational Therapy/Physical Therapy
OTSG	Office of The Army Surgeon General
PA	Physician Assistant
PAHO	Pan American Health Organization
PASBA	Patient Administration Systems and Biostatistics Activities
PERSTEMPO	Personnel Tempo
PROFIS	Professional Filler System
PT	Physical Therapy
RAF	Royal Air Force
RC	Reserve Component
RCCS	Remote Clinical Consultation System
ROEs	Rules of Engagement
RTD	Return to Duty

SOF	Special Forces
TB	Tuberculosis
TDA	Table of Distribution and Allowance
TF	Task Force
TOE	Table of Organization and Equipment
UAVs	Unmanned aerial vehicles
UN	United Nations
UNHCR	United Nations High Commission for Refugees
UNITAF	United Nations Task Force
UNMIH	United Nations Mission in Haiti
UNOSOM I/II	United Nations Operation Somalia I/II
UNPROFOR	United Nations Protection Force
URIs	Upper respiratory infections
USAID	U.S. Agency for International Development
USAMEDCOM	U.S. Army Medical Command
USAREUR	U.S. Army Europe
WAMC	Womack Army Medical Center
WHO	World Health Organization
WRAMC	Walter Reed Army Medical Center

Chapter One

INTRODUCTION

BACKGROUND

As the United States contends with the strategic uncertainty of the post–Cold War era, it must consider the role of its military forces in operations other than war (OOTW). Despite concerns about the conduct of recent UN operations and ongoing debate in Congress about limits on the U.S. role, the U.S. military can expect to be called upon to undertake such OOTW missions as peacekeeping, peace enforcement, and humanitarian assistance.

Participation in OOTW missions implies the need to provide medical support to U.S. forces or to multinational forces. In fact, one might argue that given the very nature of these operations, medical support tends to play a more central role than in combat operations. Whether the mission is to assist civilians in disaster relief, to support U.S. or multinational forces in a peacekeeping operation, to distribute medical and other supplies in a humanitarian effort, or to provide medical support in a nation-assistance program, medical support is a key component. Further, medical units often find themselves in areas where the medical infrastructure has been destroyed or there are large refugee populations. Presenting a further challenge, experience shows a strong tendency for OOTW medical missions to expand as the operation continues, with the potential to consume large amounts of scarce medical resources.

Meeting these needs will have a direct impact on the Army Medical Department (AMEDD). First, these missions come at a time when the AMEDD itself is downsizing. Several of the Army hospitals that deployed to Somalia, the United Nations Protection Force

1

(UNPROFOR), and Haiti were scheduled to deactivate upon their return. In Europe, the 7th Medical Command has recently deactivated, with U.S. European Command going from eleven to two fixed medical facilities within the past several years. Yet this theater is also the location of a number of recent OOTW and a location where several new ones may be on the horizon. The drawdown also has imposed its own set of constraints on the AMEDD, and the effects have been exacerbated by the fact that OOTW are often open-ended in nature, making it challenging to plan their medical support.

At the same time, the AMEDD has to be concerned with how to support OOTW without degrading beneficiary care. Ensuring quality medical care for the services' beneficiary population—particularly for dependents overseas—is an important concern of both the Army leadership and OASD (Health Affairs). As the current situation in Europe illustrates, the size of the U.S. force may draw down and the number of fixed facilities decline, but demand for beneficiary care does not decrease proportionally with reductions in the size of the force. How the wartime structure and peacetime structure achieve a balance in meeting these two sets of demands will continue to be an important challenge for the AMEDD.

Therefore, to minimize the impact of OOTW missions on the AMEDD's primary mission and on peacetime health care, the AMEDD needs to explicitly consider the requirements of these missions and what adjustments may be needed to accommodate them in the future.

OBJECTIVES

This report has two main purposes. First, it describes and synthesizes the lessons of recent military experience in medical support for OOTW missions. Second, it distills that experience into suggestions for improving OOTW support while minimizing the impact on the AMEDD's readiness mission and its delivery of peacetime health care.[1] We deal primarily with five key issues:

_____

[1]Although we focus on the AMEDD's role in OOTW, many of the recommendations and issues discussed herein are also applicable to the other services' military medical departments, as well as to those of U.S. coalition partners.

- How do the medical support requirements of OOTW differ from those of combat missions? What special demands are imposed by OOTW, especially when multinational forces are involved?

- How can the Army manage the inherent pressures toward open-ended expansion of the medical mission in OOTW?

- How can the AMEDD build a robust and flexible system to meet a broad range of demands associated with these operations?

- How can the Army minimize the impact of OOTW on the Army's readiness mission and its ability to deliver peacetime heath care?

- What kind of planning, education, and training may be required to better prepare the Army to support OOTW in the future?

APPROACH

To examine these issues, we use a case study approach combining both qualitative and quantitative data for selected deployments where the primary medical mission was to support U.S. forces or a multinational force. Two main operations were examined as case studies:

- The Balkans: The UN operation known as UNPROFOR and the U.S. operation known as Operation Provide Promise (OPP).

- Somalia: Operations Restore and Continue Hope (ORH and OCH).

To a lesser degree, we also examined the AMEDD's role during the operations in Haiti. This included the initial U.S.-led effort (known as Uphold/Maintain Democracy) and the follow-on UN peacekeeping mission (known as UNMIH). Because Haiti was the more recent deployment, we were unable to do as detailed an analysis of this mission as we did for the others.

Interviews were conducted with AMEDD and other military personnel who had participated in or planned these operations.[2] We fur-

---

[2]Because these personnel are typically scattered across different military treatment facilities when not deployed, this involved tracking down individuals who had participated in recent operations of interest. Further, due to the drawdown in Europe,

ther examined documents on selected deployments, as well as Army manuals and other DoD reports on such topics as OOTW and health service support in a theater of operations. These included after-action reports, field manuals, information papers, briefing charts, and numerous government and military documents (e.g., GAO reports, military pamphlets). Specific sources are cited in the text footnotes. Also, several of the medical activities (MEDDACs) and medical centers (MEDCENs) that supported these deployments by sending Professional Filler System (PROFIS) personnel provided us with both information papers and summaries on the impact of recent operations on their own activities and on patient care. Interviews with resource management officers, medical personnel, and logistics officers at these installations rounded out our information.

Specific data and methods for the operations in the Balkans and Somalia are described in the case studies presented below.

ORGANIZATION OF THE REPORT

This report is organized into several chapters. In Chapter Two, we examine how the AMEDD's wartime structure is set up for supporting traditional combat missions as a way of providing a baseline to evaluate the OOTW requirements seen in the case studies. In Chapter Three, we present the case study of the AMEDD's involvement in the Balkans, and in Chapter Four, we present the case study of Somalia. Chapter Five draws some generalizable conclusions from both case studies, and Chapter Six suggests future directions for improving AMEDD support in OOTW.

---

many of the medical personnel who had participated in UNPROFOR had already scattered to a number of different locations, complicating the interview process.

Chapter Two

TRADITIONAL ARMY MEDICAL WARTIME STRUCTURE

To understand how the medical requirements of OOTW differ from those of traditional wartime medical support, we start with a baseline of how the AMEDD is structured to support its traditional wartime mission and to provide health care services to its beneficiary population.

HOW THE ARMY HEALTH SERVICE SUPPORT SYSTEM IS ORGANIZED

The Army's health service support system is designed to be a single, integrated system that reaches from the combat zone in the theater to CONUS. The underlying idea is that the system is a continuum of care in which a soldier injured on the battlefield will be provided a full range of services, from simple first aid in the theater to more definitive care at a fixed facility within CONUS or Europe.[1]

The deployable Army medical force is made up of units and personnel from both the Active Component (AC) and the Reserve Components (RC), with 75 percent of its wartime structure being in the RC. The Army Medical Department (AMEDD) is responsible not only for supporting the Army's wartime mission, but also for maintaining the delivery of health care to its beneficiary population.

---

[1]*Health Service Support in a Theater of Operations*, FM 8-10, Headquarters, Department of the Army, 1 March 1991, p. 3-1.

The peacetime structure in CONUS primarily comprises Army hospitals—medical centers (MEDCENS) and smaller medical activities (MEDDACs)—under the command of the U.S. Army Medical Command (USAMEDCOM). These fixed facilities are referred to as the Table of Distribution and Allowance (TDA) units. The wartime structure comprises Table of Organization and Equipment (TOE) medical units assigned to combat organizations under the command of U.S. Army Forces Command (FORSCOM). The medical personnel required to staff the TOE combat units are actually assigned to the TDA units when not deployed or during peacetime. These personnel work within these Army hospitals or the peacetime structure and are designated as Professional Officer Fillers under the Professional Officer Filler System (PROFIS). PROFIS enables the AMEDD to use its wartime requirement of professionals on an everyday basis during peacetime in the delivery of health services to the Army's beneficiary population (i.e., active-duty personnel, their dependents, and retirees). Thus, under PROFIS, military health care professionals are able to maintain their clinical skills and individual level of readiness when not deployed by working within peacetime facilities.

STRUCTURE SUPPORTING THE AMEDD'S WARTIME MISSION

The process by which the AC and RC are employed to fulfill the AMEDD wartime mission is as follows. During mobilization or a contingency operation, PROFIS personnel (active-duty personnel) are pulled out of Army hospitals or military treatment facilities (MTFs) to join the combat medical unit they have been assigned to. RC TOE medical units may also be part of the deploying force. Other RC TDA hospital units or individuals are then used to backfill CONUS hospitals or hospitals within Europe that lost deploying PROFIS personnel. In addition, RC TDA medical units and individual reserve personnel can provide mobilization expansion capability, if required, by supplementing deploying active-duty medical units either as individual fillers—e.g., specialists not normally part of the wartime structure or specialists in the low-density areas of concentration (AOCs)—or as an entire unit.

Further, during wartime, the system is to shift its focus entirely to its wartime mission. In the past, transition to war meant shifting beneficiary care to the civilian sector under the Civilian Health and Medical Program of the Uniformed Services (CHAMPUS), with only active-duty service members continuing to be cared for in Army hospitals within CONUS or outside of CONUS (OCONUS). In this way, CONUS and OCONUS beds could be freed up in anticipation of Army hospitals being filled by casualties from the theater.

However, beginning with Operation Desert Storm (ODS), this policy changed. Guidance for ODS from the Army Chief of Staff (CSA) stipulated that there was to be no degradation in the care of beneficiaries as a result of this war. This policy has held ever since and is applicable to OOTW as well as to war.

This new policy fundamentally changed how the AMEDD does business, since the AMEDD now must perform its wartime mission and support OOTW while simultaneously maintaining the delivery of health care to the Army's beneficiary population. Thus, wartime and peacetime care have become inseparable, and military medical planners must now factor in how to minimize a deployment's impact on beneficiary care. Thus, the medical mission and the backfill mission are integral components of the planning process.

THE WARTIME MEDICAL MISSION: CONSERVE THE FIGHTING STRENGTH

The AMEDD's wartime mission is to support the line commander by conserving the fighting strength so that he may accomplish the military mission. The AMEDD's wartime mission then encompasses the following objectives:[2] (1) save lives; (2) clear the battlefield of casualties; (3) provide state-of-the-art care; (4) return a soldier to duty as rapidly as possible or evacuate him back to a higher echelon of care for more definitive treatment; and (5) provide the most benefit to the maximum number of personnel.

In operational medicine, the physician or combat medic will have a limited set of resources with which to save lives, treat and return a soldier to duty as far forward as possible, or stabilize and evacuate a

---

[2]*Operational Branch Concept Combat Health Support,* Directorate of Combat and Doctrine Support, U.S. Army Medical Department Center and School, Fort Sam Houston, TX, 8 September 1994, p. 6.

soldier to a more definitive level of care. In addition, the theater medical support system is designed to reduce the incidence of disease and nonbattle injury (DNBI) through good preventive medicine support to the troops.[3] Further, the medical commander must maintain enough flexibility in the theater medical system to deal with a near-overwhelming or overwhelming casualty situation.[4]

Echelons of Care

To meet these wartime needs, health service support in the theater of operations is organized into echelons of care. These echelons extend rearward throughout the theater and depend on a reliable evacuation system.

Echelon I. Echelon I—the first medical care a soldier receives—is unit-level health care that includes treatment and evacuation from the point of injury or illness to the unit's aid station. This echelon includes immediate lifesaving measures, DNBI prevention, combat stress support,[5] casualty collection, and evacuation to supporting medical treatment. At this echelon, medical care encompasses self-aid, buddy aid, combat lifesaver, combat medics, and a treatment squad (battalion aid station).[6]

Echelon II. Echelon II is division-level health service support, which includes evacuating patients from the unit-level aid stations and providing initial resuscitative treatment in division-level medical

---

[3] *Health Service Support in a Theater of Operations,* FM 8-10, Headquarters, Department of the Army, 1 March 1991, p. 1–8.

[4] A key difference between operational medicine and peacetime health care delivery is that the physician in the peacetime setting is able to draw on a broad set of clinical skills, support personnel, and equipment and supplies to provide comprehensive care to the patient. In operational wartime medicine, a physician has a limited and fixed set of resources and personnel to support the line commander, with the goal being to maximize the health benefit for the greatest number of personnel.

[5] Combat stress support is often a critical asset in OOTW missions. In these operations, coalition troops can find themselves dealing with such extreme problems as refugee populations, starving individuals, atrocities, etc. Limitations on troops' ability to respond to attacks (i.e., strict rules of engagement (ROEs)) can serve as another stressor. Although peacekeepers are "noncombatants," these troops may frequently come under sniper fire or attack.

[6] *Health Service Support in a Theater of Operations,* FM 8-10, pp. 3-3 and 3-4.

facilities. This echelon includes medical companies, support battalions, medical battalions, and forward surgical teams, as well as intratheater patient evacuation assets. At Echelon II, emergency care, including beginning resuscitation procedures, is continued. Soldiers who can be returned to duty within 24 to 72 hours are held at this echelon for treatment.

Echelon III. Echelon III is corps-level health service support, which includes evacuating patients from supported divisional and nondivisional units and providing resuscitative and hospital care. In addition, Echelon III includes providing area health service support within the corps' area to units without organic medical units. Echelon III care is provided by units such as mobile army surgical hospitals (MASH), combat support hospitals (CSH), evacuation hospitals (EVAC), and field hospitals (FH). Patients unable to survive movement over long distances receive surgical care in an Echelon III hospital. In these theater hospitals, patients receive care that will either allow them to be returned to duty or stabilized for evacuation out of the corps or out of the theater altogether.

Echelon IV. Echelon IV is communications zone-level health service support, which includes the receipt of patients evacuated from the corps. This echelon involves treating the casualty in a general hospital and other communications zone (COMMZ)–level facilities. Here, patients receive further treatment to stabilize them for their evacuation to CONUS.

Echelon V. Echelon V is the most definitive care provided to all categories of patients in CONUS and OCONUS Army hospitals. Echelon V is the CONUS-sustaining base and is where the ultimate treatment capability for patients from the theater resides, including full rehabilitative care and tertiary-level care.

Given these different echelons of care, health service support in the theater of operations is made up of a number of different elements, including hospitalization, command elements, laboratory services, medical logistics and blood management assets, evacuation assets, combat stress support, preventive medicine support, dental services, and veterinary services. All of these are components of Echelons I through V that have to be integrated to form the theater medical system. In this report we focus on the hospital units (Echelon III), but we address, where appropriate, issues specific to other types of units and other echelons of care as they affect the overall performance of the theater medical system.

PLANNING THE SUPPORT REQUIREMENTS

Although mission guidance will come down from the Joint Staff to the warfighting CINC who will develop the operational plan, the CINC designated to perform the mission will plan the support requirements with input from the supporting unified commands and the component service commands of the supported and supporting CINCs who provide the forces. Upon receipt of the mission, the supported CINC, with the assistance of assigned service component commands, will initiate the planning process for developing a concept of operations and related tasks; this concept determines the forces needed to accomplish the mission. As part of this process, the medical planners will undertake their mission estimate and analysis based on the mission statement and mission guidance provided by the JCS. Given the specific mission, this process may involve coordinating health service support requirements with allied or other friendly forces involved in an operation.

Planning factors for the medical mission itself will include: (a) the number of troops to support, (b) the population at risk, (c) the expected casualty or combat intensity rates, (d) the expected DNBI rates, (e) bed availability, (f) the expected admission rates, and (g) the theater evacuation policy.[7] The theater evacuation policy will state the maximum period that casualties who are not expected to return to duty (RTD) may be held within the theater for treatment before being evacuated out.[8] This policy is established by the Secretary of Defense upon advice of the JCS and recommendations of the designated CINC.

---

[7]Planning of the medical support is based on the doctrinal employment of medical units and the organizational capability of these units. In addition, for UN missions there isoften a force cap imposed that sets in advance the total size of the U.S. force. The force cap in turn highly constrains the size of the medical component of the U.S. force.

[8]*Doctrine for Health Service Support in Joint Operations,* Joint Pub 4-02, 26 April 1995.

In addition, planning takes into account what is available locally, whether the U.S. military will be going into a region with an underdeveloped medical infrastructure or one that has been completely destroyed, and what is available in surrounding countries in terms of medical facilities and other resources.[9]   All these factors will determine the support requirements, the evacuation policy, and how the echelons of care get set up.

Planning for the medical mission also will take into account a unit's readiness level, where it is in the training cycle, whether it has trained for a particular type of operation, and the experience of its commander and his staff. It also takes into account whether RC units or reserve medical personnel may be used for a given operation. Moreover, planning takes into account how to minimize the impact of a deployment on operating tempo (OPTEMPO) and personnel tempo (PERSTEMPO). Finally, planning takes into account the backfill requirements and how to minimize the impact of a deployment on beneficiary care.

All of these items factor into the selection of medical units, the number and mix of medical personnel required, and the planning of subsequent rotations.

ASSUMPTIONS UNDERLYING HEALTH SERVICE SUPPORT IN A THEATER OF OPERATIONS

Given the design of the U.S. military's health service support system, a series of explicit and implicit assumptions underlie the planning process in terms of the nature of the medical mission, how the theater medical system or the medical structure will be organized, what patient populations will be served, and what the demand for medical services will be. Underlying all of this is the fact that the system itself is designed specifically to support the U.S. military's wartime mission. This in turn will have implications for planning in terms of the medical intelligence requirements, specialty mix of medical personnel and type of units required, preventive medicine support needed,

---

[9]That is, how far away you are from the nearest tertiary care facility in a neighboring country. Tactical planners also consider what might be the largest force that gets engaged.

equipment and supplies required, and organization of the echelons of care for a given operation. As this report will show, these assumptions break down in a number of key areas when the mission is to support OOTW or peacetime contingency operations involving a multinational force. The important assumptions underlying health service support in a theater of operations are summarized below.

Assumptions About the Medical Mission

- The U.S. Army's health service support system is designed to support specifically U.S. troops versus foreign contingents, civilians, or children. As discussed below, this has important implications for the nature of the patient population expected to be served.

- The U.S. military is accustomed to establishing the medical policy for an operation.[10] This means that historically:

  — In general, the United States does not expect to have to deal with differences among coalition partners in medical objectives or medical policies for a given operation. It also does not expect to have to account for differences in the quality of deploying forces' medical assets or variation in levels of physical readiness among their troops.

  — The United States does not expect to have to support or coordinate with civilian relief organizations or local community medical providers in an operation.

  — In the past no repatriation policy has been necessary, since the U.S. military expects all forces to abide by the evacuation policy set for the mission.

- A fundamental tenet underlying the health service support system is that the United States "takes care of its own" (i.e., only U.S. military personnel will provide care to U.S. soldiers). So if

---

[10]Historically, the U.S. military has had the opportunity in combined exercises to train with other medical forces and their combat units and has supported combined operations in World War II, Korea, and Vietnam. In addition, the NATO War Surgery Handbook is a coalition–agreed-upon text that outlines echelons of care and other clinical issues. However, whether units and personnel are currently trained to anticipate the impact of combined operations is a different question.

U.S. soldiers are sent to fight for their country, then they are assured of receiving the highest quality of medical care available within the United States.

Assumptions About the Medical Structure

- Health service support will be a single, integrated system that reaches from the forward area of a combat zone as far rearward as the patient's condition requires, including to CONUS.[11]

- The different echelons of care will be connected, allowing for the uninterrupted care and treatment of the wounded, injured, or sick.[12]

- Fixed facilities will exist at Echelons IV and V to which patients can be evacuated from the theater.

- The system is geared toward stabilizing and evacuating patients from the theater when necessary to a more definitive level of care.

- A viable and timely evacuation system exists with dedicated MEDEVAC aircraft and personnel assigned to the mission.

- A viable medical logistics supply system exists that is based on U.S. FDA standards.

- The U.S. military's standards of quality of medical care and of equipment, units, personnel, and training will be adhered to.

Assumptions About the Nature of the Patient Population

- Troops will represent a healthy, young adult population (and predominantly male).

- Troops will have a high level of medical and dental readiness, minimizing the number of chronic and acute medical conditions that may require treatment in the theater of operations.

---

[11]*Doctrine for Health Service Support in Joint Operations,* Joint Pub 4-02, 16 April 1995, p. 1-6.

[12]Ibid.

- Troops will have good preventive medicine support throughout the course of deployment and during the predeployment phases.

Assumptions About the Demand for Services

- The demand for medical services will tend to be primarily for trauma and surgical care, since good preventive medicine support and a high level of physical readiness of troops will serve to minimize disease and nonbattle injuries (DNBIs) requiring treatment in the theater itself.

- The range of diseases expected to require treatment in the theater will be limited to naturally occurring and common infectious diseases (such as upper respiratory infections) and to diseases endemic to a particular region.

Chapter Three

UNPROFOR AND OPERATION PROVIDE PROMISE, THE BALKANS: A CASE STUDY OF THE MEDICAL MISSION

INTRODUCTION

Coalition operations pose a unique set of challenges in providing the medical support for a mission in which troops are drawn from a variety of nations to create a multinational force. In these operations, coalition partners may bring in varying levels of quality of medical assets, as well as a wide range of diseases. Differences in standards of care, definitions of echelons of care, and in the level and quality of health care within a coalition soldier's own country may also influence patient care and evacuation decisions in a theater of operations.

Such was certainly the case for the United Nations Protection Force (UNPROFOR) in the Balkans, which was a coalition of UN and NATO forces for a peacekeeping mission initially established on 21 February 1992. In November 1992, UNPROFOR initially comprised units from 31 countries organized into 15 active battalions, with forces consisting of 23,000 UNPROFOR troops in the region, including the civilian employees and contract personnel associated with UNPROFOR and NATO employees and officers; by March 1994, those forces had grown to more than 40,000.[1] This force was to be an interim arrangement to create the conditions of peace and security required for negotiating a settlement of the Yugoslav crisis.

Operation Provide Promise (OPP) stood up 1 February 1993, with Joint Task Force Provide Promise (JTF-PP) established to consolidate

---

[1]At its peak, the U.S. hospital served a UN military population of more than 47,300.

the oversight of a variety of U.S. missions in the former Yugoslavia, including: (a) command of all U.S. forces operating in support of UN operations in the Balkans; (b) air-land transportation and airdrops of humanitarian relief supplies into Bosnia-Herzegovina; (c) provision of medical support to UNPROFOR troops; (d) detecting, monitoring, and reporting activities along the border of Serbia and the former Yugoslav Republic of Macedonia (FYROM); and (e) conducting reconnaissance using unmanned aerial vehicles (UAVs) in support of UN, NATO, and U.S. operations.[2]

The U.S. medical mission during UNPROFOR and Operation Provide Promise was to provide Echelon III support to UN peacekeeping forces. This mission was jointly shared by all three services, with each undertaking one or more rotations of the U.S. military hospital in Zagreb, Croatia. During the time period covered by this case study, the Army undertook the first two rotations, followed by the Air Force and then the Navy, which undertook the fourth rotation. See Table 3.1 for a list of the service rotations.[3]

Table 3.1

U.S. Hospitals and Rotations for UNPROFOR

U.S. Hospitals	Rotation Dates
212th MASH (Army)	15 November 1992—27 April 1993
502nd MASH (Army)	28 April 1993—08 October 1993
48th ATH (Air Force)	09 October 1993—16 March 1994
Fleet Hospital 6 (Navy)	17 March 1994—29 August 1994

SOURCE: Data from briefing, "Operation Provide Promise, a Nursing Perspective," CAPT Nancy Owen, Director of Nursing Services, Fleet Hospital 6.

---

[2]"Joint Task Force Provide Promise Deactivates," 1 February 1996, Defenselink News Release.

[3]This report covers the experience of only the first four rotations of U.S. hospitals during UNPROFOR. In August 1994, Fleet Hospital 5 assumed the medical mission. In March 1995, the Air Force assumed the mission from the Navy and has had it continuously until early 1996, when the U.S. medical mission for UNPROFOR and Provide Promise officially ended.

In addition, the U.S. hospital in Zagreb, Croatia augmented the medical support of U.S. forces participating in the ongoing UN peacekeeping operation Able Sentry in Macedonia.

This chapter provides a case study of the U.S. role in providing the medical support mission for coalition forces participating in UN-PROFOR and OPP, with a specific focus on the Army's medical role.[4]

We start by examining the medical mission statement and the nature of the medical structure. We next turn to the demand for medical services during the operation, looking specifically at what services were demanded and at the nature of the patient population to be served. Finally, we examine how well the requirements met this demand.

In all cases, the intent here is to describe first what was expected or assumed in each of these areas. Then we examine what actually happened in each of these areas, showing how it varied from conditions at the outset.

MEDICAL MISSION STATEMENT AND MEDICAL STRUCTURE

The medical mission statement and the medical structure for UN-PROFOR and OPP are intertwined. The mission statement is what medical support U.S. forces are tasked to perform within the coalition, while the medical structure is how the coalition sets up assets and policies to execute that mission. Medical structure issues center around how the echelons of care are set up in-theater and what the evacuation and repatriation policies are. These assets and policies are driven by the differing medical assets and policies of the coalition partners, as well as by the UN's policy for medical logistics and the state of the local medical infrastructure.

---

[4]The focus in this report is on the Army's medical experience during UNPROFOR and OPP. But because this was a joint mission, we also discuss in some detail the experiences of the Air Force's and Navy's hospitals to fully understand how the medical mission evolved during this coalition operation.

Medical Mission Statement at the Outset

Initially, the medical mission during UNPROFOR was limited to providing support to truck convoys delivering humanitarian relief supplies to Bosnia-Herzegovina. During this initial phase, UNPRO-FOR units provided their own Echelons I and II care, with the UN purchasing Echelon III care from civilian contractors for these units. However, by July 1992 the situation in Bosnia-Herzegovina had begun to heat up, and in October 1992, European Command (EUCOM) received the order for the U.S. military to undertake the medical mission of providing Echelon III care to UNPROFOR troops. By November 1992, the U.S. Army had established a Mobile Army Surgical Hospital (MASH) at Camp Pleso on the outskirts of Zagreb, Croatia, as part of UNPROFOR.

In providing Echelon III care, the initial medical mission statement called for providing hospitalization and comprehensive care to all UN forces for up to 30 days (i.e., an evacuation policy of 30 days), including the treatment of UN civilian employees and contract personnel associated with UNPROFOR.[5]

Medical Structure at the Outset

Echelons of care, evacuation and repatriation policies, and the contributions of coalition partners' assets. As mentioned above, the United States had responsibility for Echelon III care during the operation. The rest of the system was set up as follows. Echelon I care—defined by U.S. standards to include battalion aid station, combat medic, combat lifesaver, combat stress support, and buddy aid—was to be the responsibility of each contingent; this included treatment and evacuation capabilities from the point of injury or illness to the unit's aid station and the evacuation of patients from the field to Echelon II.

At the outset of the operation, Echelon I care ranged widely among the various contingents—all the way from a medic with a first aid kit to the U.S. Army's definition. In addition, some troops lacked key as-

---

[5]This operation was unusual also in that U.S. medical units are seldom directly responsible for force protection (e.g., security, safety issues).

sets considered integral to the overall mission, such as preventive medicine support and combat stress capabilities. In UNPROFOR, many developing countries in particular lacked preventive medicine assets. In addition, their predeployment screening and preparations tended to be inadequate or nonexistent (e.g., no immunizations, no chemoprophylaxis or medications to protect their troops against infectious diseases, no health screening of troops prior to deployment).[6] Most contingents also lacked combat stress support and were thus unable to identify or treat mental health problems in the field.

The British were initially tasked for Echelon II care. Their responsibilities were to include intratheater evacuation (ambulance and some air assets),[7] the liaison function connecting Echelons I and III care, and the provision of a significant preventive medicine activity. The British also initially had overall command and control of the medical support for UNPROFOR. In addition, the French were to provide air evacuation for Bosnia; in Croatia, UN contractors were used to provide this support.

Echelon II care also included the forward surgical teams (FSTs) provided by the various contingents, with most battalions having an FST attached.[8] The FSTs varied in size, composition, and quality of medical care provided (like the Echelon I medical assets of the various contingents). In addition, the FSTs were not evenly distributed across the various sectors.[9]

Echelons IV and V care involved continued health care outside the theater and was intended to be a national responsibility, as was the repatriation of troops (i.e., each country was to be responsible for the transport of its own soldiers out of the theater).

---

[6]These differences were related in part to the level of development of a country's health care system, cultural differences, differences in the amount of value placed on the armed forces, and for some of the poorer nations, an inability to afford these types of assets.

[7]Interview with CAPT Johnson, Commander, Fleet Hospital 6, 22 February 1995.

[8]Echelon II can include medical companies, support battalions, and/or forward surgical teams, in addition to intratheater patient evacuation assets.

[9]For example, during the fourth rotation, the southwest and southeast sectors (where the heaviest casualties were occurring) were not covered by FSTs.

What Happened to the Medical Mission During the Operation

As we look across the various rotations during the operation, we see how the medical mission began to evolve over time; this evolution was driven by directions from above, as well as by changes in the provision of the echelons of care and by evacuation and repatriation problems.

During the initial rotation, the Army's 212th MASH unit followed closely the written mission guidance of providing Echelon III care to UNPROFOR forces and to UN and NATO personnel. Early command pressure was for the task force to stay within the mission parameters. Task Force (TF) 212, the initial American medical component of this operation, thus did not get into providing refugee care, with the only civilians treated being either UN employees, NATO employees or officers, or contract personnel.[10] Neither did TF 212 send medical personnel out into the various sectors.[11] In addition, initially the 212th MASH was the only Echelon III hospital in the theater.

Further, the task force had established a liaison with the local community hospitals, contracting for use of certain medical equipment (e.g., CAT scanner), as well as setting up a contingency plan whereby local hospitals had agreed to take on patients from the 212th MASH in a mass-casualty situation in order to free up beds and Army medical personnel.[12]

The Army's 502nd MASH unit, which took over the medical mission in April 1993, continued a range of activities and policies similar to those established by TF 212.

---

[10]The U.S. military was not authorized to care per se for contract personnel, even though they were treated during this operation. The TF enforced the policy of not treating refugees for several reasons: (a) UNPROFOR forces were to remain neutral; (b) the situation was one of multiple adversaries (i.e., Croats, Bosnians, and Serbs) and constantly shifting ethnic alliances; (c) the fact that TF 212 was surrounded by a large refugee population, the needs of which a 60-bed hospital could not begin to meet. In addition, the medical staff were kept from volunteering during their off-duty time. Interview with COL Gregg Stevens, Commander, TF 212.

[11]TF 212 was the initial American medical component of UNPROFOR and OPP. On 1 February 1993, Joint Task Force Provide Promise (Forward) stood up, taking on oversight of all U.S. missions in the former Yugoslavia.

[12]TF 212 brought in Class 8 (medical supplies) and Class 9 (repair parts) supplies, but was able to purchase most other supplies locally.

When the Air Force's 48th Air Transportable Hospital (ATH)[13] took over in October 1994, two events occurred that led to a significant change and expansion of the medical mission.

First, the British medical battalion—responsible for Echelon II capabilities—pulled out.[14] The British departure meant the loss of much of UNPROFOR's assets for intratheater ground transportation of patients, the loss of the liaison function connecting Echelons I and III care, and the loss of the significant preventive medicine activity the British had performed. With the departure of the British medical battalion, there were no longer dedicated helicopters or vehicles for MEDEVAC or ground transportation of patients. Although the Norwegians would subsequently be tasked to assume Echelon II activities, they did not come on-line until much later, during the fourth rotation when the Navy's Fleet Hospital 6 had already replaced the Air Force's 48th ATH in Zagreb.[15] In addition, the Norwegians were located in Tuzla, which meant that their Echelon II capability was located far from where most of the casualties were occurring.

As a result, the Air Force (and later the Navy) undertook the job of filling in the void in Echelon II assets and assuming the liaison function for Echelon II.[16] The Air Force did so by providing a transport team to assist in the aeromedical evacuation of UNPROFOR person-

---

[13]The ATH can provide Echelons II–III care. The ATH is configured in increments of 14, 25, 50, and 90 beds. The 50-bed and over ATHs are capable of providing a full range of medical services, including an operating room, and may be augmented by a Hospital Surgical Expansion Package (HSEP).

[14]The departure of the British medical battalion was a political decision. Initially, Britain had agreed to provide Echelon II assets for the first rotation. But as the second rotation drew near, the UN still had not moved toward replacing the British medical battalion with other patient transportation assets. The British reluctantly agreed to provide Echelon II for the second rotation to give the UN time to find a replacement. By the third rotation the UN still had not come through as promised, and so the British Echelon II assets were pulled out, although the battalion and medics themselves wanted to stay on and continue their mission. Interview with COL Robert Leitch, British medical liaison officer, OTSG.

[15]In addition, the Pakistanis had been tasked by the UN to take on the Echelon II function, but these troops arrived in-theater untrained and unequipped to do so. Even though the troops received extensive training and equipping, by the end of the fourth rotation they were still not ready to take over this function.

[16]From the briefing "48th Medical Group Air Transportable Hospital, Camp Pleso, Republic of Croatia," by MAJ Patrick Throop, Chief, Medical Logistics Flight.

nel and by establishing a sector liaison program in which medical teams were to be sent out to the various sectors. Sector activities included assessing the medical assets of other contingency forces and the quality of those assets, observing first-hand the various sectors' medical problems and hygiene and environmental conditions, and educating the various coalition forces on the type of medical support available to them from the 48th ATH.

The second factor that changed the medical mission was growing interest by the UN, the U.S. State Department, and the JTF Command for the Air Force to begin treating refugee children and adults at the ATH.[17] A memo from the Chief of Staff dated 5 January 1994 instructed the AF 48th ATH to set aside five acute-care beds for the treatment of both refugee children and adults. The UN further sought assistance from the Air Force in providing medical help to UN personnel and their children, especially for UN personnel working in the more dangerous regions of the Balkans. In particular, there was a demand for psychiatric services and combat stress support.

As a result, the Air Force established liaisons with refugee camps (e.g., Varazdin) and began taking children for elective surgery and dental care, as well as adults for medical/surgical evaluation. The Air Force also extended hospital visits and established a medical liaison with the refugee camp at Cakovec. The 48th ATH further began coordinating with the UNHCR, WHO, UNPROFOR, and various relief agencies in treating and evacuating refugee casualties.

The operation parameters for the 48th ATH entailed being able to: (1) provide up to 40 surgical operations within a 72-hour window before requiring augmentation; (2) provide liaison and coordination with nongovernmental organizations (NGOs) in the use of the MASH for treating and evacuating refugee casualties; and (3) insofar as the primary mission allowed and in coordination with the UN Force Chief Medical Officer and Joint Task Force Provide Promise (forward) Commander (JTFPP(FWD)-CO), (a) assist with the medical evacuation of UNPROFOR personnel, (b) provide on-site medical and

---

[17]Calls for doing so came from a number of different quarters, including U.S. Ambassador to the UN Madelaine Albright and Secretary of State Warren Christopher. In response to such requests for refugee care, the Joint Staff asked JTF-PP(F) to do a study on requirements for establishing a pediatric trauma ward in November 1993.

technical assistance to UNPROFOR medical officers and units and educational assistance in the management of cardiac and traumatic emergencies, and (c) provide planning, technical, and physical assistance to the UN Force Chief Medical Officer.[18]

Importantly, the Air Force assumed responsibility for both filling in Echelon II and providing refugee care during the third rotation without written mission guidance and with only verbal approval from higher authority to extend the scope of the medical mission. Although the Air Force commander continually sought written guidance, this was not obtained during the six-month rotation.

When the Navy Fleet Hospital 6 took over the medical mission on 17 March 1994, it took over the Echelon II activities begun by the Air Force and extended them. This included creating seven MEDEVAC teams to be on 24-hour call who were also trained to convert helicopters to litter-bearing aircraft.[19] The Navy continued the sector liaison program established by the Air Force but broadened it, increasing the number of sector visits. Such sector visits were conducted for a variety of reasons. For example, doctors visited each of the sectors as part of the Echelon II mission, corpsmen were sent to Tuzla to train on the Norwegians' armored ambulance evacuation capability, the hospital commander visited Tuzla to coordinate with the Norwegians on addressing the hole in Echelon II capabilities, a surgeon was sent to backfill the Norwegian hospital while their medical personnel rotated in, independent duty corpsmen were lent several times to the Canadians to fill in gaps in their medical assets, the commander sent a Navy Seabee Detachment to Split and Sarajevo to fix equipment, etc., a Navy psychiatrist was sent to Sarajevo to set up a program for individuals continually under fire and to provide group therapy for UN peacekeepers who had been held hostage by the Serbs, and a preventive medicine officer was sent to Sarajevo to assess public health conditions. (This officer was able to identify a major problem in the city's water purification system.) Navy physi-

---

[18]From the briefing "48th Medical Group Air Transportable Hospital, Camp Pleso, Republic of Croatia," by MAJ Patrick Throop, Chief, Medical Logistics Flight.

[19]These teams comprised a physician, nurse, and several corpsmen (some had physician assistants).

cians were also each assigned to a sector and were to be responsible for entry-level evacuations and for the coordination of care within it.

Another important change in the medical mission occurred early in this rotation: There was a significant increase in the number of trauma and trauma-related patients as a result of land mine injuries, which led to a change in the medical support requirements as well as to Fleet Hospital 6 taking on Echelon IV care as well.

Echelon IV care comprised the following. The Fleet Hospital 6 had established a close working relationship with the local hospitals, appointing an officer to serve as liaison. Because Zagreb's hospitals were quite capable by Western standards and had available sophisticated medical equipment (e.g., CAT scan) and expertise (e.g., neurosurgical consults and an operating capability), the Navy was able to arrange access to these capabilities and so provide Echelon IV care to coalition patients. As it turned out, some of the world's experts in the treatment of complex mine injuries were located at the University of Zagreb. The Fleet Hospital 6 staff and the university medical staff at one point held a combined symposium on complex mine injuries.

In addition to assuming Echelon IV care and filling the void in Echelon II assets, the Navy continued to serve as the primary interface with UNPROFOR, UNHCR, the U.S. embassies, and the relief agencies. Like the Army, however, the Navy provided very little refugee care; any treatment of civilians was for those who were associated with either the UN or NATO.

Finally, the Navy was able to augment the assets of U.S. forces participating in Able Sentry, the ongoing peacekeeping mission in Macedonia. For example, the Fleet Hospital 6 sent orthopedic teams monthly to assist the local physicians and implemented an aggressive physical therapy and education program for these forces. In addition, it sent a Catholic chaplain to Macedonia and made several visits to screen and put on classes on preventive treatment for U.S. forces.

In summary, the medical mission throughout UNPROFOR kept expanding in response to changing support requirements. Initially, the Army remained fairly restrictive in its activities, closely following written mission guidance. But by the third rotation, increased pressure to take on refugee care and the hole in Echelon II assets forced the Air Force to expand its activities beyond the scope of the original medical mission. The Navy similarly continued the activities the Air Force had begun to keep the in-theater medical system intact. In addition, melting snows coupled with a cease-fire led to more land mine injuries and thus an increase in the number of trauma patients seen by the Navy's fleet hospital and to the assumption of Echelon IV care during the fourth rotation.

What Happened to Echelons of Care During the Operation

As mentioned earlier, there was a wide variety in the quality of assets and medical care within the various echelons of care at the outset. During the operation, poor quality of care and inappropriate care at Echelons I and II meant that the United States, in some instances, ended up treating coalition patients with unnecessary complications. Further, delays in transporting patients to the U.S. hospital resulted in some cases in a worsening of the patient's medical condition. Thus, U.S. medical personnel sometimes found themselves having to undo what the Echelon I assets or the FSTs had done out in the field. For example, the 212th MASH had a 21-year-old Eastearn European soldier who arrived at the hospital with a badly swollen face and a life-threatening gum infection. The Army's maxillofacial surgeon ended up having to pull a number of the soldier's teeth and treat him with massive doses of antibiotics to save his life. In another case, an Eastern European soldier with a chest wound had had a chest tube inserted by one of the FSTs, but he was not transported to the 212th MASH until two days later. Upon arrival, his chest tube had become clogged and infected. It took an Army surgeon four days of constant vigil to get the soldier cleaned out and stabilized.

Also as mentioned earlier, many countries did not have combat stress support assets. Consequently, during UNPROFOR, coalition troops would often seek out combat stress support from the U.S. hospital, which provided it although such support was not within the scope of the original medical mission. In general, although Army, Air Force, and Navy combat stress personnel did what they could to fill in this void,[20] they also faced some difficult challenges, including linguistic and cultural barriers.

Inadequate health screening and predeployment preparations among some of the contingents[21] meant that some soldiers were in relatively poor health status when they arrived in the theater, with some troops bringing in serious infectious diseases such as malaria and tuberculosis, as well as diseases not endemic to the region.[22] Of the infectious diseases, tuberculosis was particularly problematic, with an estimated 40 percent prevalence among soldiers from the former Soviet republics.[23]

Overall, there was wide variability in the medical and dental readiness of UNPROFOR forces. Troops from West European countries, Canada, and Australia were comparable to the United States in their level of medical readiness, although the dental readiness of some was not up to U.S. standards. Troops from some developing countries (e.g., former Soviet Republics, Poland) showed more variability in their level of medical readiness and almost no dental readiness. Troops from other developing countries (e.g., those from African countries, the Pakistanis, the Nepalese) showed the most variability in their level of medical readiness, with dental readiness being virtually nil.

Evacuation and repatriation problems during the operation. The departure of the British medical battalion left a hole in Echelon II as-

---

[20]For example, during the operation, five UN peacekeepers who were held hostage by the Serbs were continuously threatened with death throughout their ordeal. A Navy psychiatrist was sent to Sarajevo to set up group therapy for these hostages as well as a program for civilians living under constant attack.

[21]The fact that some countries obtain their peacekeeping forces through advertisements suggests that very little, if any, screening is done.

[22]In general, soldiers from developing countries tended to be in poorer health and to have more variable health status than soldiers from West European countries, Canada, or Australia.

[23]Other examples: the Navy fleet hospital treated some advanced cases of tuberculosis (e.g., patients with lung abscesses). Hepatitis B and C were also of great concern. Other infectious diseases included HIV, chicken pox, mumps, typhoid, and measles. In addition, U.S. hospitals saw individuals with diabetes as well as chronic heart conditions. The U.S. hospital also treated some patients with diseases one would not normally expect to have to treat in the theater, such as Hodgkin's disease.

sets. This meant that the intratheater evacuation of patients became difficult and circuitous. Specifically, there were no longer dedicated helicopters or vehicles for MEDEVAC or ground transportation of patients between Echelons I and III. Although the UN contracted out for local ground transportation and for air evacuation, using borrowed vehicles and ambulances, some of these vehicles were not specifically designed for patient transport.

In addition, the UN had contracted out MEDEVAC services with a KLM subsidiary helicopter company (IRA). Some of these helicopters, however, were not specifically set up for MEDEVAC, did not have flight crews trained in the aeroevacuation of patients, and lacked medical personnel who could provide the hospital staff on the ground with the type of information needed to arrange the transport of a patient.

Further, an evacuation request from the FSTs had to be transmitted by nonmedical personnel through several layers of UN bureaucracy, further complicating the evacuation process. For example, the FSTs would send a request for patient transport to the UN, which would then notify the U.S. hospital via the JTF air liaison officer.[24] The hospital staff, in turn, would have to go through the JTF air liaison officer (nonmedical officer) to make arrangements to transport the patient. The fact that intratheater evacuation was being handled by civilians and nonmedical personnel meant that the U.S. hospital at times would receive calls to send a medical team to evacuate a patient but would not receive adequate information on the patient's condition, the number of patients to be transported, or the type of medical personnel and resources needed. The Fleet Hospital 6 staff tried to get around this by talking directly to the FST physicians in the field. This was not always possible, however, since the Serbs were continually interrupting the communication lines. The Serbs also controlled the air space, limiting the periods during which a MEDEVAC mission could be flown.

---

[24]The way intratheater evacuation is supposed to work under U.S. standards is that the higher level picks up from the lower level of care. For example, an Echelon III hospital would typically have an ambulance company attached to it and would be responsible for picking up patients from the FSTs or from a medical company (Echelon II).

Repatriation, or intertheater evacuation of patients, was to be a national responsibility. In general, the Canadians, Australians, and the West European countries were able to pick up their injured soldiers and transport them home. However, the repatriation of soldiers from the developing countries (e.g., the former Soviet Republics, African nations, and Middle Eastern and Far Eastern countries) was more of a problem. Many of these countries lacked an air evacuation capability. Instead, special arrangements had to be made to retrieve an injured soldier, which often meant a significant delay in patient pick-up from the Echelon III hospital.[25]

As a result, each of the U.S. military hospitals ended up holding on to some coalition patients far longer than they would normally expect to or than was medically necessary. During the fourth rotation, the repatriation of coalition soldiers from developing countries averaged around 2–4 weeks, whereas evacuation out of the theater of soldiers from the more developed countries averaged one week. Although the UN had published an evacuation policy, it became situational, depending on the ability or willingness of different contingency forces to repatriate their own soldiers. During UNPROFOR, the U.S. Army patient administration officers quickly learned to start the paperwork to have a coalition patient evacuated out of the theater as soon as he entered the hospital. Such difficulties in repatriating soldiers from developing countries created problems for the United States in terms of patient tracking and for the medical staff in terms of ethical and treatment dilemmas, since for some patients there was no one to transfer their care to.[26]

The first rotation supplies another example of how concerns about the availability and quality of health care within a soldier's own country led the United States to hold onto certain patients longer

---

[25]The way repatriation was supposed to work was that the U.S. hospital would submit a request to the Force Surgeon, who submitted a request to the UN commander, who then turned to the individual contingency's battalion commander to request arrangements be made to send a soldier back home.

[26]For the most part, the medical personnel we interviewed did not feel that there was a lot of unnecessary care-seeking behavior among the coalition forces. Any anger that the medical staff did express was related more to the fact that because some soldiers were from countries with poor health care systems that could not adequately meet their needs, the United States ended up bearing responsibility for these soldiers in filling in what was missing in their own country.

than it would normally expect to in order to ensure that the soldier received appropriate care. The Army's 212th MASH took on the care of four Russian soldiers who were amputees and had been languishing in a community hospital in Zagreb. These soldiers had received inadequate nursing care, and as a result their muscles had begun to atrophy. The Russian ambassador knew that if these soldiers were returned home they would probably be unable to obtain prostheses and also that there was a good prosthetics manufacturer in Zagreb. The ambassador intervened on their behalf, requesting that the 212th MASH get them back into good medical condition and house them until they could be fitted with prostheses. However, fitting these soldiers with prostheses and getting them through the initial stages of rehabilitation took time. The 212th MASH ended up keeping these soldiers a maximum of 89 days, a stay almost three times longer than the mission's 30-day evacuation policy. The Air Force and Navy had similar cases during their rotations. As this example illustrates, the level of development of a country's health care system was a contributing factor in the United States assuming Echelon IV care for some patients.

Medical Logistics at the Outset

Going into the operation, DoD planners assumed the U.S. task force could rely on the UN logistical supply system. Maintaining the blood supply was another critical medical logistics issue for UNPROFOR, with the various contingents intended to be responsible for their own blood supply.

What Happened to Medical Logistics During the Operation

The assumption that the United States could rely on the UN logistical supply system turned out to be unrealistic during the operation. When UNPROFOR started, the UN supply system for this operation was nonexistent. After it was set up, it was slow, taking anywhere from 4 to 6 weeks to fill requests.[27] All the U.S. hospitals instead

---

[27]Another reason why U.S. medical units could not utilize the UN medical logistics system was FDA requirements. Because the UN accepts donated items, there was no quality control over those items; in addition, no single set of standards was adhered to.

established a petty cash fund from which they could purchase supplies locally rather than go through the UN supply system. In general, the U.S. hospitals ended up going through U.S. channels to obtain most of their supplies and other support requirements.[28]

Maintaining the blood supply was a critical issue throughout UN-PROFOR. Although each contingent was supposed to provide its own, in reality only the United States and West European countries had the capability to do this. In addition, U.S. medical personnel's ability to tap into the civilian blood supply was limited and dependent on whether the local populace itself was dealing with a combat casualty situation. If so, then civilian casualties might mean a high demand for blood. In such instances, one would have to be able to bring in outside sources of blood. EUCOM implemented the first frozen blood supply program in a field environment during the fourth rotation. In addition to cultural sensitivities about who would receive whose blood, there were concerns about procedures for screening the blood supply for HIV-related viruses. For example, some European countries have a lower HIV rate than the United States, and for this reason alone some forces were wary of receiving U.S. blood. On the other hand, the United States had similar concerns of its own. During this operation, the U.S. policy was to use only U.S. blood to treat both U.S. personnel and coalition soldiers, since some countries do not routinely screen for certain HIV-related viruses.[29]

DEMAND FOR SERVICES

In looking at the demand for services during the operation, we examine the combination of two factors: (1) the population to be served, which centers around the mix between U.S. and other forces and between military and civilian personnel, as well as age and gender dif-

---

Recall that one of the assumptions in the AMEDD's wartime structure was that medical supplies would meet FDA requirements.

[28]This also suggests that most of the supplies either came out of the CINC or the individual Services' budgets.

[29]From the beginning of this operation, the policy was to use U.S. blood first from Landstuhl, then take blood from U.S. soldiers, then others, and as a last resort use the local blood supply only if the need was critical. Interview with COL Stevens, CDR, TF 212.

ferences and the level of medical and dental readiness of the troops; and (2) patient demand, which centers around differences in demand for trauma versus primary care, the amount of disease and the type of medical conditions and injuries requiring treatment, and changes in the level of demand over the course of this deployment. Again, we first examine the expectations at the outset and then what actually occurred during the operation.

Expectations of Populations Served at the Outset

In terms of populations to be served, the expectation going in as established by the medical mission statement, was that the U.S. hospital would primarily be treating coalition forces from other countries, including some UN civilian employees and contract personnel associated with UNPROFOR. The mission statement excluded foreign civilians and refugees from the population to be served. There was also an implicit assumption that the hospital would be primarily dealing with troops with a high level of medical and dental readiness.

Populations Served During the Operation

Table 3.2 shows the number of admissions and outpatient visits by patient category for the two Army hospitals. On the inpatient side, the Army primarily took care of foreign military personnel (UNPROFOR forces) and NATO personnel. For the 212th MASH, 78 percent of its admissions were in these two patient categories, whereas only 14 percent of admissions were U.S. personnel.[30] Foreign civilians comprised the remaining 8 percent of admissions. These proportions stayed roughly the same for the 502nd MASH.[31]

In terms of outpatient visits, approximately 38 percent of the 212th MASH's outpatient visits were by U.S. personnel and 28 percent by

---

[30]Outpatient visits and admissions by U.S. personnel were primarily by U.S. military personnel. A small number of U.S. civilians were also in-theater and so are included in this category (e.g., visiting dignitaries, embassy personnel, intelligence personnel).

[31]Throughout, the proportion of admissions that were U.S. personnel remained relatively low: 14 percent (45/333) of admissions for the 212th MASH, 12 percent (34/288) of admissions for the 502nd MASH, 18 percent (58/323) of admissions for the 48th ATH, and 9 percent (33/353) of admissions for the Fleet Hospital 6.

Table 3.2

Comparison of Total Number of Admissions and Outpatient Visits by Patient Category for the Army's Rotations in UNPROFOR

	212th MASH		502nd MASH	
	Admissions	Outpatient Visits	Admissions	Outpatient Visits
U.S. personnel	45	1,387	34	2,173
Foreign military	126	1,038	100	1,105
NATO employees/ officers/UN	134	879	127	1,247
Foreign civilians	28	362	27	611
Total	333	3,666	288	5,136

SOURCE: Data are from the Directorate of Patient Administration Systems and Biostatistics Activities (PASBA), AMEDD Center and School, Fort Sam Houston.

foreign military personnel. NATO employees and military personnel were continually coming and going in the theater to evaluate NATO's requirements and to set up the no-fly zone. NATO personnel and UN employees[32] accounted for a quarter of the outpatient visits and foreign civilians[33] for 10 percent of the visits.

During the Army's second rotation, the distribution of outpatient visits across the four patient categories stayed approximately the same. The only difference between the first and second rotations was a decrease in the percent of outpatient visits accounted for by foreign military personnel (from 28 to 22 percent).

The reality of the medical condition of patients to be served during the operation differed significantly from initial expectations. As mentioned earlier, coalition forces from a number of countries lacked the medical and dental readiness that is generally assumed for U.S. forces. In addition, preventive medicine support in-theater was

[32]The majority of NATO personnel in the theater were either NATO or UN employees; very few were NATO military officers.

[33]Foreign civilian personnel in UNPROFOR were primarily UN contract personnel, generally Croats, who provided various services such as food, laundry, waste disposal, etc. to the JTF and the UNPROFOR headquarters in Zagreb.

often lacking. As a result, U.S. hospitals treated a wide variety of acute and chronic medical conditions, as well as such serious infectious diseases as tuberculosis.[34] In addition, demand for emergent dental care was relatively high throughout the course of UNPROFOR due to the low levels of dental readiness.[35]

Because many UN, NATO, and contract personnel were women and because some coalition forces had a large number of female soldiers, there was also a high demand for ob/gyn care in the theater. In addition, requests to assist in the evacuation and treatment of refugee children led to the need for some pediatric services.

Further, because many countries relied heavily on reservists and civilian contract personnel, these troops tended to be older, to be in poorer health status, and to have a wider range of acute and chronic medical conditions than soldiers from countries who used primarily active-duty soldiers. As a result, for example, the U.S. medical staff had to evaluate some older soldiers and civilians for such conditions as acute chest pain during UNPROFOR.

The end result of this wide mix of patient groups was that the U.S. hospitals were pushed in the direction of providing a broader range of services, since their patient population more closely resembled that of a community hospital, as opposed to what a military hospital would expect to see in a theater of operations. To illustrate how these differences in patient groups translated into the nature of the patient population seen by the U.S. hospitals, we describe the Navy's Fleet Hospital 6 inpatient experience:

- First, over half of the admissions to the Navy hospital were for trauma or trauma-related injuries, (including complex mine injuries to the extremities, multiple shrapnel wounds, head trauma, burns to the extremities, etc.). There were also orthopedic injuries, some of which were sports-related.

---

[34]Because a number of countries indiscriminately prescribe antibiotics, U.S. medical personnel also saw drug-resistant strains, including tuberculosis.

[35]Very little of the demand for dental services during UNPROFOR was for nonemergent problems. In fact, some soldiers had so many dental problems that the U.S. military dentist would treat the most critical problem and within a month or so the soldier would be in for the others.

- Second were gastrointestinal problems (e.g., GI bleeding, peptic ulcer, gastroenteritis, appendicitis, abdominal pain), most of which are not uncommon in a young adult male population. There were also a number of stones—kidney stones, urethral stones—requiring treatment.

- Third, there were a number of dental procedures (e.g., wisdom tooth extractions, odontectomies, abscesses, etc.), which for the most part required only short hospital stays of one or two days.

- Fourth, there were chest pains of various sorts requiring inpatient evaluation (e.g., possible myocardial infarction, atypical chest pain, etc.), as well as a few other chronic medical problems including diabetes and chronic otitis media.

- Fifth, there were infectious diseases such as tuberculosis, malaria, chicken pox, hepatitis, upper respiratory tract infections, and pneumonia.

- Sixth, there were a few rare events (e.g., Hodgkin's disease, brain tumor) and several psychiatric cases including depression, suicidal ideation, psychosis, and alcohol intoxication.

Expectations of Patient Demand at the Outset

The initial expectation in terms of patient demand at the outset was that the U.S. Echelon III hospital would primarily be treating diseases and relatively few injuries, since most of the combat casualties were expected to occur among the civilian population, not among peacekeeping forces. Recall that the initial medical mission statement called for providing hospitalization and comprehensive care for up to 30 days, which excluded the provision of long-term rehabilitative care or more definitive therapy for patients in the recuperative phase (i.e., Echelon IV care).

In contrast, the forward surgical teams (FSTs) were expected to see the majority of emergency trauma patients in the theater. However, because of the wide variability in the quality of medical assets across UNPROFOR troops, this meant for some soldiers that if they were able to make their way to the U.S. hospital and survive any delays in intratheater evacuation or substandard care that might be provided at Echelons I or II, then by the time they reached the U.S. hospital they probably would be stabilized and require more reconstructive or rehabilitative care than trauma care.

In general, soldiers from West European countries tended to remain in the care of the United States only as long as was medically necessary; in some instances they would bypass the U.S. hospital in Zagreb altogether and be flown directly out of the theater to a fixed facility in their country.

On the other hand, soldiers from developing countries would tend to remain in the care of the U.S. hospital somewhat longer. These patients were also at greatest risk of developing complications, experiencing delays in evacuation, and receiving poor quality of care in the field. Coupled with repatriation problems, these soldiers tended to be more resource intensive to treat and to be more likely to require Echelon IV care. In addition, some patients who no longer needed medical attention still had to be housed on a minimal-care ward until they could be repatriated or returned to their unit.

Below we examine how these differences translated in terms of patient demand and length of stay differences across the various troops.

Patient Demand During the Operation

Table 3.3 shows the total number of admissions and outpatient visits by rotation during this operation. Not surprisingly, the demand for outpatient services increased as the size of the UN force increased from the initial 23,000 troops to more than 40,000 by the fourth rotation. In contrast, the number of admissions for the four rotations remained fairly constant throughout the deployment.

What did change on the inpatient side, however, was the proportion of admissions that were injury-related (trauma) versus disease-related. Before the fourth rotation, two-thirds of all hospital admissions were disease-related. However, during the Navy's watch, half of the fleet hospital's admissions were now trauma-related, suggesting that the fleet hospital took on more resource-intensive patients than had the previous three hospitals.

Table 3.3

Comparison of Total Number of Admissions and Outpatient Visits by Rotation for UNPROFOR

U.S. Hospitals	Number of Outpatient Visits	Number of Admissions	Proportion of Disease to Injuries
212th MASH (Army)	4,454	338	67/33
502nd MASH (Army)	4,715	313	63/37
48th ATH (Air Force)	6,610	323	64/36
Fleet Hospital 6 (Navy)	9,131	353	48/52

NOTE: The total outpatient visits and admissions listed for the 212th and 502nd MASH units differ from those reported in Table 3.2. The reason for the discrepancy is the use of two different data sources. Table 3.2 indicated an increase of about 1,500 visits, whereas Table 3.3 shows no increase in outpatient visits between the first and second rotations. It was necessary to use two different data sources because only the PASBA data allowed for a breakdown of the utilization pattern across patient categories. In addition, because the PASBA data gave us information only on the Army's rotations, data on the third and fourth rotations had to be obtained from briefing charts. Both sources, however, indicate an overall trend of increasing outpatient visits over time. Further, both sources indicate that the admission rate remained fairly constant over time.

SOURCE: Data from briefing charts: "Operation Provide Promise, a Nursing Perspective," CAPT Nancy Owen, Fleet Hospital 6's Director of Nursing Services. The proportional distribution column refers to admissions only.

This shift in the proportion of trauma-related admissions corresponds with changes in the operation that took place at that time. As noted above, by the spring of 1994, melting snows and the cease-fire led to an increase in the number of trauma patients with complex mine injuries. There were also important length-of-stay differences across patient categories. Table 3.4 shows the average length of stay by patient category for each of the Army hospitals.

Comparing length of stay for different patient groups, we see that the pattern differs for the two Army hospitals. Overall, average length of stay was twice as long during the first rotation (212th MASH) than during the second rotation (i.e., 7.2 days versus 3.7 days). Within the different patient categories, foreign military personnel had the longest average length of stay (10.3 days) and U.S. personnel the shortest (3.2 days) during the first rotation. The relative rank ordering of the four patient groups in terms of length of stay remained the

Table 3.4

Average Length of Stay by Patient Category for the Army's Rotations in UNPROFOR

	212th MASH		502nd MASH	
	Admissions	Average Length of Stay	Admissions	Average Length of Stay
U.S. Personnel	45	3.2	34	1.8
Foreign military	126	10.3	100	4.9
NATO employees/ officers/UN	134	5.8	127	3.5
Foreign civilians	28	7.3	27	2.4
Total	333	7.2	288	3.7

SOURCE: Data are from the Directorate of Patient Administration Systems and Biostatistics Activities (PASBA), AMEDD Center and School, Fort Sam Houston.

same during the second rotation, although length of stay dropped from 1 to 5 days on average within each category.[36] The shorter stays of U.S. personnel may be largely attributable to the fact that these personnel were primarily support personnel and were not near any of the heavy fighting.

Previously, we noted the wide variability among coalition forces in their medical and dental readiness. Given this variability, the average length of stay of foreign military personnel shown in Table 3.4 is somewhat misleading, since this category combines forces with different lengths of stay. To examine these differences, we used inpatient data from the fourth rotation to compare length of stay across the various contingents.[37] In Table 3.5, troops are grouped by country of origin.

---

[36]The drop in average length of stay for the foreign military category during the second rotation may reflect a learning curve in that the 502nd MASH had the benefit of the 212th MASH's experience in terms of dealing with the repatriation process.

[37]We use data from the Navy to illustrate length-of-stay differences, because the Army's data could not be disaggregated by contingency force.

Table 3.5

Comparison of Average Length of Stay of UNPROFOR Troops by Country for the Fourth Rotation

Contingency Force	Number of Outpatient Visits	Number of Admissions	Average LOS (days)
United States	1,989	33	3.8
Netherlands	507	12	2.0
Finland	193	1	2.0
Sweden	186	15	2.6
Canada	357	26	5.4
France	509	21	5.8
Norway	270	13	5.8
Britain	805	34	6.4
Slovak Republic	102	8	5.6
Argentina	171	8	4.4
Egypt	119	5	4.7
Jordan	588	38	8.7
Nepal	139	12	6.1
Pakistan	279	21	6.1
Kenya	221	10	13.8
Poland	163	18	6.9
Russia	309	24	18.7
Ukraine	178	9	15.9
Total	7,085	308	7.8

SOURCE: Data from CAPT Carlisle, Navy Fleet Hospital 6.

NOTE: A few contingents were missing length of stay data and so are not shown. Also, throughout this operation, the composition of the force changed. Therefore, the countries shown here represent only a partial list of the nations who contributed forces during the entire deployment.

Although there is a "small-numbers" problem for some forces, in general we see that the United States and the West European countries tended to have the shortest average lengths of stay. Argentina and Egypt fell into the middle range. Beginning with Jordan, we start to see longer average LOS across the remaining contingency forces, ranging from 6.1 to 18.7 days. The coalition forces with the highest average LOS also tended to have more outlier cases.[38] The former Soviet republics had some of the longest-staying patients. For example, a Russian soldier with an admitting diagnosis of tuberculosis had a LOS of 41 days; two Russian soldiers with mine injuries had stays of 100 and 134 days respectively; and a Ukrainian soldier with Hodgkin's disease had a LOS of 48 days.

Although variation in length of stay across UNPROFOR troops likely reflects differences in physical readiness, in quality of their medical assets, as well as in medical need, some of the longer-staying patients were not necessarily more resource-intensive to treat. For example, the Russian soldiers who were amputees did require rehabilitative care, but they also required that the MASH unit house them until they could be fitted with prostheses. In other cases, delays in repatriation accounted for some of the additional days.

Were U.S. hospitals kept busy during UNPROFOR? During the initial rotation, patient demand was lower than had been anticipated. The reasons for this were severalfold. As mentioned above, the Army closely followed written mission guidance and was more restrictive in the scope of its activities (e.g., it did not undertake refugee care). In addition, during the Army's two rotations, Echelon II was working reasonably well. Further, the U.S. hospital was never located near the mainstream of the casualty movement although the potential was there with the hospital's location at Pleso airfield.

The Air Force added on Echelon II care, sector visits, and refugee care to the primary mission of providing hospital care to UNPROFOR forces. The Navy extended the Air Force's activities and, as discussed above, saw a greater proportion of trauma patients. In addition, over time there was an increase in the total number of UNPROFOR troops requiring medical support. Nonetheless, at no point were any of the hospitals overwhelmed by patient demand in the theater or by combat casualties.

---

[38]For example, a Jordanian soldier with land mine injuries and multiple shrapnel wounds had a LOS of 56 days; a Kenyan soldier with mass in the right upper quadrant had a LOS of 26 days; and a Pakistani soldier with an admitting diagnosis of fever of unknown origin had a LOS of 28 days, to name a few.

REQUIREMENTS TO MEET DEMAND FOR SERVICES

Given the differences in types of patients deployed to the theater and the types of services they required, we next examine how well the requirements sent to the theater met these demands.

Medical Requirements at the Outset

During the initial rotation by the Army, the decision was made to send a MASH unit, which could more readily be broken down into its component parts, rather than to deploy a combat support hospital. The 212th MASH out of Wiesbaden, Germany initially was configured to have twelve ICU beds, two 20-bed intermediate-care units, and eight minimal-care beds (holding units).[39] The MASH unit, normally designed to provide emergency care only, was tailored to include a wide range of services (e.g., physical therapy, surgery, internal medicine, dentistry, emergency, etc.).[40] Again, as mentioned above, the operation parameters called for the 212th MASH to be able to provide comprehensive care and hospitalization to all UN forces for up to 30 days.

Medical Requirements During the Operation

Although advance assessment had paid off in terms of the hospital's configuration and the range of services required, the sizing was off.[41] As mentioned earlier, the demand for services was relatively low during the initial phases of UNPROFOR. Thus, as the first rotation progressed, one of the intermediate-care units was eventually stepped down to a minimal-care unit. Low patient demand also led the TF 212 to send 43 of the 397 originally deployed personnel (257 were directly affiliated with the Army's MASH unit) back to their parent units, with the option of recalling them to the theater later on, if necessary.

---

[39]The initial requirement called for a 60-bed capability with 30 medical and 30 surgical beds.

[40]A MASH unit is typically designed for 72-hour emergency care and then evacuate.

[41]Note that typically one will plan for the worst-case scenario and then tailor back upon arrival in-theater once an assessment of the situation can be made.

All subsequent rotations fell in on the 212th MASH's hospital and equipment. In late April 1993, the Army's 502nd MASH unit replaced the 212th in the theater. Although the 502nd MASH made some minor modifications to the hospital and adjustments to the mix of personnel brought into the theater, basically the setup was quite similar during the two Army rotations.

In October 1993, the Air Force's 48th Medical Group assumed the medical mission.[42] It comprised 142 AF medical personnel, of whom 99 were from the Royal Air Force Base (RAF) Lakeheath, Suffolk, England, and 40 were from other U.S. military hospitals and clinics within Europe.[43] Similar to the two Army hospitals, the 48th ATH was tasked to provide a 60-bed surgical hospital with 30 beds for minimal-care patients.

The 48th ATH similarly was set up to provide a full range of services, including (a) an eight-bed intensive care unit; (b) two isolation tents; (c) two medical/surgical wards (one of which was later utilized as a classroom and to house adults accompanying refugee children); (d) 24-hour emergency services capability; (e) a pharmacy; (f) extensive postinjury physiotherapy and follow-up orthopedic care; (g) radiology capability;[44] (h) one two-tent section (four bunks) for psychiatric treatment; (i) dental services; and (j) a variety of other services (e.g., environmental health, medical logistics, patient administration, medical food service, and communications).

The Air Force was tasked to develop a plan to establish a pediatric ward for refugee children and ultimately set aside five acute-care beds for the treatment of refugee adults and children. As noted earlier, the other significant requirement during the third rotation was

---

[42]The JTF Commander was COL Watkins; the Hospital Commander of the 48th ATH was COL Steve Jennings.

[43]The mix of medical personnel included three general surgeons, one orthopedic surgeon, one anesthesiologist and four nurse anesthetists, one psychiatrist and one mental health technician, two family practitioners, two internists, two physician assistants, twenty staff nurses and the chief nurse, one physical therapist and one PT technician, one general dentist, one oral surgeon and six dental technicians, one pharmacist and three pharmacy technicians, one environmental health officer, one laboratory officer, four MSC officers, and miscellaneous other enlisted personnel including laboratory, x-ray, cardiopulmonary, etc. technicians. The commander was also a surgeon.

[44]The 48th ATH also contracted locally for MRI and CT studies.

for the Air Force to train its medical personnel to conduct MEDEVAC missions and to take on the sector liaison function.

The Navy assumed the medical mission in March 1994. The Fleet Hospital 6 also was tasked to provide Echelon III care for all UNPROFOR personnel. Its capabilities included 24-hour emergent and nonemergent care, a 60-bed hospital (24 acute-care beds, 6 isolation beds, and 30 minimal-care beds), physical therapy, and respiratory therapy service. Fleet Hospital 6's medical staff included 171 personnel, with a total of 313 personnel comprising JTF (FWD) Provide Promise.

The Fleet Hospital 6 initially had tailored what it brought into the theater and its mix of providers based on the previous three rotations' patient demand and support requirements. However, with the change in patient mix and the requirement for Echelon IV care, the Fleet Hospital 6 ended up making the following adjustments:[45]

- Due to the increase in trauma patients, the number of orthopedic surgeons was increased to two and the number of general surgeons was reduced by one.[46]

- The Fleet Hospital 6 greatly expanded its physical therapy department. The Navy also implemented an aggressive physical therapy and preventive treatment program. Accordingly, two-thirds of the physical therapy cases were referred by primary care physicians (or via the ER) rather than by an orthopedic surgeon. The Navy did so in an attempt to cut down on the number of serious orthopedic injuries.[47] This meant bringing in a physical therapist and two PT technicians.

- About half of the Navy's patients were females, so a general practitioner with training in ob/gyn care was brought in and a cubicle set aside specifically for this purpose.

---

[45]Interview with CAPT Johnson, Commander, Fleet Hospital 6, 22 February 1995.

[46]In addition, one of the anesthesiologists was sent home, so the fleet hospital operated with one anesthesiologist and two nurse anesthetists.

[47]The preventive physical therapy program enabled the hospital to rehabilitate soldiers quickly (e.g., within a week) and return them to their units rather than evacuating them out of the theater for treatment.

- Due to the high demand for emergency dental care, two general dentists and one oral surgeon were deployed.

In addition, the large number of trauma patients and the attending requirement for rehabilitative services required the fleet hospital to be innovative in improvising traction capabilities, etc. The medical staff further had to quickly become familiar with the treatment of complex mine injuries. Like the Air Force medical staff, the Navy's medical staff were trained for MEDEVAC missions and were assigned sector responsibilities.

In addition, all four hospitals had to contend with the following support requirements. First, the above-discussed problems in repatriating soldiers and returning them to their unit required a minimal holding unit capability. In addition, beds were needed for buddies of injured soldiers who came to the U.S. hospital to serve as the patient's translator. Further, soldiers who required such services as emergency dental care but did not need to be hospitalized still had to be housed. Adults who accompanied refugee children also required housing. The U.S. hospital at one point housed for an extended period several orphaned children.

Second, there was a need for an isolation capability in-theater. Achieving an isolation capability in a tent environment, however, is difficult to do. At times there were patients with a variety of different contagious diseases all housed on the same ward.[48]

Third, the large number of female patients meant a requirement for an ob/gyn setup (e.g., a cubicle) to do gynecological exams, gynecological medical supplies and equipment, and a physician trained in ob/gyn care. Each hospital had to improvise to accommodate this type of patient demand. For example, the Navy brought over a general practitioner who had received extra training in this area and could bring his own instruments.

Finally, the need for translators and linguistic support was high throughout this operation. For example, initially over 31 countries

---

[48]In addition, U.S. medical personnel were exposed to a variety of serious infectious diseases. Three Navy medical personnel, for example, showed positive skin tests for exposure to tuberculosis upon their return from UNPROFOR.

were participating in UNPROFOR. Although the U.S. hospital was located near UNPROFOR headquarters in Zagreb, the UN was not always able to assist with translation. Some nationalities, such as the French and Jordanians, might send another soldier to accompany a patient and serve as his translator. Other troops such as the Russians would drop a patient off and then depart. Not only did this cause problems in communicating with the patient about his treatment, it also caused trouble tracking down his unit when the patient was to be discharged or critical treatment decisions needed to be made. Each of the hospitals had a different method for meeting their linguistic requirements. The 212th MASH, for example, had several Army personnel fluent in a number of Balkan languages whom it relied upon for translation. The Air Force hired a Croatian physician for its outpatient clinic who served as the interface between the 48th ATH and the local community hospitals. The U.S. Navy field tested a mechanical translation device during this operation.

In summary, throughout the course of UNPROFOR, sizing of the medical support was never much of a problem. In fact, as shown earlier, the number of patients admitted to each hospital remained fairly constant throughout this operation. Although the number of outpatient visits steadily increased over time, this alone was not as significant a determining factor of the medical support requirements as were other variables.

The real problem was that the mission itself was quite fluid in terms of the types of patients the hospital would end up treating and the range of activities U.S. medical personnel would be required to take on to keep the in-theater medical system going. In addition, throughout this operation there were security concerns for U.S. personnel.

Further, the medical issues that arose during UNPROFOR became more and more complex over time, including refugee care, coordination of refugee patients' evacuation and treatment with the UN, UNHCR, and various other relief agencies, assumption of Echelon IV care, problems in the repatriation of soldiers, gaps in Echelon II, and treatment of complex mine injuries. In addition, there were the above-discussed differences in medical readiness among the UNPROFOR troops and in the quality of medical assets in-theater. Combined, these factors meant that it was difficult to predict at any one time what mix of personnel, units, supplies, and equipment was needed in the theater.

OVERALL CASE-SPECIFIC OBSERVATIONS

When we look across the experience of the medical mission in the Balkans, a number of observations emerge. On the whole, these center around the unique problems associated with working with the UN and with coalition forces.

Problems Working with the UN

A number of the problems that occurred during UNPROFOR were the result of multiple layers of command and control. During UNPROFOR, for example, there were Joint Staff orders, UN orders, EUCOM orders, and requests being made by the State Department. In the case of refugee care, at times there were conflicting requests. For example, the Air Force's 48th ATH technically belonged to the UN, which requested that this hospital provide treatment to refugee adults. At the same time, the U.S. State Department had requested that the 48th ATH take on the care of refugee children, although technically the State Department had no real authority over the hospital.

UNPROFOR headquarters was not always in agreement with the concerns of the U.S. task force. For example, the UN commander and staff did not respond to requests by the U.S. Army about hospital security at Camp Pleso.[49] Underlying this were differences in perceptions of the level of threat and degree of emphasis on force protection.

The UN Force Chief Medical Officer is the senior medical officer in the theater with oversight over all the other force medical officers. This relationship is more than a technical one in that the UN Force Chief Medical Officer also controls the funds and can put "fences" around how the other medical officers go about their mission. For example, the UN Force Chief Medical Officer approves supplies, sets forth the medical ROEs, and enforces compliance with them. The

_____

[49]As a result, the TF 212 ended up initiating its own base defense plan.

problem, however, is that in UN operations the political realities are such that individual decisions made by coalition partners may affect the functioning of the theater medical system and at times affect the decisions made by the UN Force Chief Medical Officer. For example, during UNPROFOR the U.S. chain of command repeatedly vetoed requests by the UN Force Chief Medical Officer. In other instances, the United States was not allowed to fly into high-risk areas for MED-EVAC missions. As seen also during UNPROFOR, the decision of the British to withdraw their medical battalion impacted Echelon II. Thus, in UN operations it is political realities in terms of what each country can and cannot do that determine significantly how well the theater medical system itself functions.[50]

The various layers of bureaucracy and unclear chain of command also made it difficult to accomplish certain tasks during UNPROFOR, such as purchasing supplies. The UN also had cumbersome reporting requirements. For UNPROFOR, it had set up a complex system to distribute funds. TF 212 had to assign extra personnel just to monitor UN funds and comply with UN reporting requirements.[51]

In terms of logistics, planners were unrealistic in assuming that the U.S. hospital could rely on the UN medical logistics system. The UN system is highly constrained by funding. Further, that system is slow, taking anywhere from 4 to 6 weeks to fill requests. Part of the problem during UNPROFOR was that the U.S. force needed to get used to working within the UN bureaucracy and the process by which requests were approved and processed. Another aspect of the problem was the fact that U.S. medical units are required to comply with FDA standards, and so were unable to utilize some UN medical supplies.

Other problems arose because of mismatched tour lengths. UN tours are normally one year, whereas U.S. tours are typically 179 days. It takes some time to figure out how the UN system is supposed to work and then to work the system. This meant that the various U.S. task forces would just get established, work out the

---

[50]Interview with COL Lester Martinez-Lopez, UNMIH Force Chief Medical Officer, August 1996.

[51]This was in addition to the personnel required to handle U.S. internal reporting and accounting requirements (e.g., payroll, petty cash, local contracts).

problems in the system, and start to run smoothly for a relatively short period before the new rotation came in.

Problems Working with Other Coalition Partners

Throughout the case study, we see numerous examples of problems arising out of differences between the United States and its coalition partners. These problems centered around differing medical policies, differing levels of assets, differing standards of care, and differing levels of physical readiness.

Differing medical policies. In terms of medical policies, countries in UNPROFOR had different views of how broad the medical mission should be. Countries with a long history of undertaking peacekeeping and humanitarian relief missions—such as Norway, Canada, the Netherlands, and Sweden—tend to define their medical mission more broadly. For example, the Norwegians normally expect to get involved with the host country and local community in providing medical care and public health services and in rebuilding the medical infrastructure. The Norwegian hospital in Tuzla, for example, not only provided care to UNPROFOR troops, but also worked with the local hospitals.

By contrast, the U.S. policy was to provide medical support to the U.S. troops and other coalition forces involved in the given relief effort and not get involved in providing refugee care or in rebuilding of the medical infrastructure, except on a very limited basis. During UNPROFOR, the U.S. medical units adhered for the most part to this primary mission.

Differences in coalition partners' medical policies can set up unrealistic expectations as to the U.S. medical mission and complicated the interactions of the U.S. military and other countries' medical teams (as well as civilian health care providers, government officials, and NGOs). For example, during UNPROFOR, the press played up the fact that the U.S. hospital did not provide refugee care.

Other coalition partners' medical policies also had the potential to affect the morale of U.S. medical personnel. During UNPROFOR, the Norwegians were kept busy treating civilians and working in the refugee camps and with the local hospitals around Tuzla. In contrast, because patient demand during UNPROFOR was relatively low and U.S. medical personnel were restricted in their activities (e.g., not allowed to volunteer while off duty to work in refugee camps or to assist relief agencies), U.S. medical personnel were at various times underutilized.

There were also varying levels of professionalism and commitment to the medical mission across the various contingents during UNPROFOR. Some contingents' contract personnel or reserve volunteers stayed for the entire rotation, while others left before their tour of duty was completed. U.S. military medical personnel were there for the long haul, tended to be held to more rigid professional standards, and often worked and lived under more restrictive conditions (e.g., 12-hour shifts).

In addition, some countries had no prohibitions on the use of alcohol by their troops, and others had less stringent policies than those of the United States.[52] Consequently, some forces had a high rate of alcohol use in the theater.

Differing levels of assets. Beyond problems caused by differing medical policies, there were also problems due to the fact that coalition partners brought varying levels of quality and types of medical assets to the table.

Medical assets in-theater were often highly variable in type and quality as a result of being drawn from a number of different nations, each with varying levels of development of their national health care systems. Some medical units came into the theater inadequately equipped, supplied, and trained. Some troops lacked key assets considered integral to the mission (e.g., preventive medicine support). Wide variability in medical assets across contingents, however, could not always be attributed just to differences among nations in the level of development of their country's health care systems. Even among the more developed countries, there were impor-

---

[52]Consumption of alcohol also has the potential to create internal security problems for a task force. For example, during the initial rotation of the hospital, Camp Pleso had present troops from seven different countries, all of whom had to share the same enclosed compound, even though historically some had been enemies. Consumption of alcohol, therefore, had the potential to lead to volatile situations. Mitigating such problems was one of the concerns senior officers faced.

tant differences in definitions of echelons of care. For example, the Norwegian hospital in Tuzla was termed an Echelon III facility, but by U.S. standards the hospital lacked key capabilities and so was considered more of an "Echelon II.5" facility. There were also important discrepancies in terms of what various coalition partners agreed to provide versus what they ultimately delivered.[53]

Some contingents also had difficulty keeping their medical activity going during UNPROFOR. For OOTW missions, a number of countries rely on civilians, reservists, and/or volunteers. The Norwegian hospital, for example, was staffed by contract physicians. However, more and more countries have been experiencing recruitment and retention problems because of the increased frequency of OOTW deployments.[54] As a result, during UNPROFOR the Navy ended up loaning medical personnel several times to other forces because of their difficulties in recruiting medical personnel for this particular mission.

Variability in the quality and type of medical assets across UNPROFOR forces meant that the United States ended up plugging in the holes in the medical assets of the other coalition partners. As one JTF commander noted, the U.S. hospital saw part of its job to be to identify and fill in those gaps.

Differing standards of care. Differences in standards of care among countries both at home and in the theater also influenced patient care and evacuation decisions in UNPROFOR. Quality of care among the various troops' medical teams varied widely, with those from developing countries tending to have lower standards of care than the United States.

---

[53]A good illustration of this comes from the U.S. experience during Operation Provide Comfort in northern Iraq. For this humanitarian relief effort, a number of countries had signed on to provide medical assets. Some countries, though, did not deliver in the final analysis, and others delivered less than what had been originally promised. For example, of the 9–10 countries involved in this humanitarian relief effort, one West European country sent a field hospital but no medical personnel to staff it, another sent a medical team but no equipment or supplies, and a third sent a battalion aid station instead of a promised field hospital.

[54]Particularly those with a long history of participation in OOTW missions, e.g., Canada, Norway, United Kingdom.

In addition, extreme differences in quality of medical care among some coalition countries were often an incentive for a soldier to be left by his countrymen in the care of the U.S. military hospital. Caring for some coalition soldiers had the potential to lead to resourcing problems if they drew heavily on critical supplies or required prolonged and intensive nursing and physician care.

On the other hand, some coalition forces from Western Europe or the other highly developed countries participating in UNPROFOR (e.g., the French and Canadians) did not always want the United States to treat their soldiers. Instead, some preferred that the U.S. hospital only house their soldiers until arrangements could be made to evacuate them out of the theater. The reasons for this were twofold: (1) like the United States, some countries have a policy of "taking care of their own"; (2) in addition, some contingents were wary of receiving U.S. blood out of fear of contamination with the HIV virus.[55] Thus, in some instances, a soldier who was transported to the U.S. hospital was simply held at his country's request until he could be picked up. This proved to be frustrating to U.S. military physicians, since at times they would have patients in their care they knew they could help but were instructed not to.

---

[55]Some West European countries have a much lower HIV rate than that of the U.S. population. The Croats in particular had a very low infection rate in their population and were especially fearful of receiving contaminated blood. Interview with CAPT Johnson, Commander, Fleet Hospital 6, 22 February 1995.

Chapter Four

OPERATIONS RESTORE HOPE AND CONTINUE HOPE, SOMALIA: A CASE STUDY OF THE MEDICAL MISSION

> Somalia was a nation divided and torn apart by a civil war. . . . Bandits ruled the major lines of communications. . . . All supply lines were blocked by roadblocks to extort "tolls," and ambushes were a way of life. . . . Twenty-four hours a day [U.S.] soldiers lived with the threat of being shot at, having a hand grenade thrown at them or receiving indirect fire attacks.[1]

INTRODUCTION

Somalia was a case of the AMEDD doing its combat mission in an operations other than war context. OOTW can range from well-defined missions with clear-cut, limited objectives to less well-defined missions with open-ended endpoints; Somalia was one of the latter.[2]

In Somalia, the nature of the medical mission was in terms of peaks and valleys in that the demand for medical services overall was relatively low and primarily for routine care, yet this mission was punctuated with periods of combat. Planning the medical support for such operations can be a difficult challenge, as commanders need to be able to respond to the worst-case scenario and yet make the best

---

[1]U.S. Army Forces, Somalia—10th Mountain Division (LI) *After Action Report Summary*, dated 2 June 1993, p. 19.

[2]In this case study, we focus specifically on the medical mission. For an overview of ORH and OCH, we refer the reader to the two after-action reports on Somalia completed by the Center for Army Lessons Learned (CALL): *Operation Restore Hope Lessons Learned Report*, 3 December 1992–4 May 1993; and *U.S. Army Operations in Support of UNOSOM II*, 4 May 1993–31 March 1994.

use of their medical assets. In addition, because patient demand tends to be relatively low in these operations, the medical staff may be underutilized at times. As a result, there is a natural tendency in OOTW to want to use any excess medical capacity for purposes that may go beyond the original medical mission.

Unlike in the Balkans, where U.S. forces were part of a coalition, in Operations Restore Hope (ORH) and Continue Hope (OCH) the United States was the primary actor in the humanitarian and peace enforcement mission undertaken by the UN in Somalia. Prior to ORH and OCH (in July and August 1992), the UN had undertaken a limited peacekeeping effort in Somalia—UN Operations Somalia, UNOSOM I—with a small contingent of Pakistanis having been sent as monitors of the March 1992 cease-fire. Around the same time, the U.S. effort in Operation Provide Relief (OPR) began, involving the airlift of humanitarian relief supplies from Mombassa, Kenya, for the NGOs operating in Somalia. However, by September 1992 it was clear that conditions were rapidly deteriorating in Somalia and that security for the relief convoys had become critical and would require a larger force than had originally been anticipated.

OPR led directly into ORH, which officially started in January 1993, with planning of this operation having begun in mid-November 1992.[3] ORH was the U.S. component of the UN's humanitarian (and peacekeeping) effort—United Task Force, UNITAF. Operating under a UN mandate, UNITAF's mission was to secure relief operations in the assigned Humanitarian Relief Sectors, with the United States responsible for four of the nine of them. The ultimate goal was to transfer all responsibilities of the mission over to UNOSOM II by May 1993. UNITAF evolved into a joint and combined task force led by the United States under UN auspices. The Commander-in-Chief, Central Command (CINC, CENTCOM) was tasked for this mission.

During UNITAF, there was also pressure from the UN for the task force to expand the original relief convoys and security mission to include disarmament and to establish a presence in the northern section of the country. The U.S.-led task force strongly resisted this expansion of the UNITAF mission but eventually did undertake some limited disarmament.

In early May 1993, UNITAF transferred responsibilities to UNOSOM II, and OCH began (the U.S. component of UNOSOM II). Starting in January 1994, the withdrawal of U.S. forces from the theater began, with a March 1994 deadline set for removing all but a small contingent from Somalia.

During the seventeen months of the Somalia deployment in ORH and OCH, the AMEDD had two major operational components: (1) medical units organic to the 10th Infantry Division (Mountain) and Task Force Kismayo; and (2) medical units included as part of the Joint Task Force Support Command-Somalia (JTFSC-S). During that time there were three rotations of medical units into the theater. Table 4.1 lists the medical units deployed to Somalia for each rotation by home base and by medical group or task force. The medical units listed are in addition to the organic medical assets belonging to the 10th Mountain Division.

As in the previous chapter, here we examine the medical support requirements for Somalia, looking first at how the medical mission evolved over the course of this deployment and at how patient demand changed over time. We then compare how well demand matched the medical support provided and the specialty mix of providers in the theater and discuss the implications for planning.

MEDICAL MISSION STATEMENT AND MEDICAL SUPPORT

Medical Mission Statement and Medical Support at the Outset

The U.S. medical mission at the outset was to provide comprehensive care to all U.S. forces involved in the security and humanitarian mission and to provide limited support to other coalition forces in the theater (i.e., on an emergency-only basis). More specifically, the initial medical mission was twofold: (1) deploy Army medical units required to support the deploying force and then tailor back once in-theater; and (2) provide treatment of combat casualties as well as routine DNBI for Army and other U.S. forces in Somalia.

Table 4.1

Army Medical Units Deployed to Somalia as Part of ORH and OCH

1st Rotation
(ORH: 9 December 1992 to 1 May 1993)

62nd Medical Group (Fort Lewis, WA)

86th Evacuation Hospital (Fort Campbell, KY)
32nd Medical Logistics Battalion (Ft Bragg, NC)
159th Medical Company (Air Ambulance) (Germany)
514th Medical Company (Ambulance) (Ft Lewis, WA)
423rd Medical Company (Clearing) (Ft Lewis, WA)
61st Medical Detachment (Sanitation) (Ft Campbell, KY)
224th Medical Detachment (Sanitation) (Ft Lewis, WA)
227th Medical Detachment (Epidemiology) (Ft Lewis, WA)
485th Medical Detachment (Entomology) (Ft Polk, LA)
555th Medical Detachment (Forward Surgery) (Ft Hood, TX)
73rd Veterinary Detachment (Ft Lewis, WA)
248th Veterinary Detachment(Ft Bragg, NC)
257th Dental Detachment (Ft Bragg, NC)
528th Combat Stress Control Detachment (Ft Bragg, NC)

2nd Rotation
(OCH: 1 May 1993 to 15 August 1993

42nd Medical Task Force

42nd Field Hospital (FH) (Ft Knox, KY)
147th Medical Logistics Battalion (Ft Sam Houston, TX)
61st Area Support Medical Battalion (Ft Hood, TX)
45th Medical Company (Air Ambulance) (Germany)
105th Medical Detachment (Sanitation) (Ft Polk, LA)
248th Veterinary Detachment (Ft Bragg, NC)
528th Combat Stress Control Detachment (Ft Bragg, NC)

3rd Rotation
(OCH: 15 August 1993 to 31 March 1994)

46th Medical Task Force

46th Combat Support Hospital (Ft Devens, MA)
32nd Medical Logistics Battalion (Ft Bragg, NC)
61st Area Support Medical Battalion (Ft Bragg, NC)
82nd Medical Company (Air Ambulance) (Ft Riley, KS)
926th Preventive Medicine Detachment (Ft Benning, GA)
248th Veterinary Detachment (Ft Bragg, NC)
47th Forward Support Medical Company (Ft Bragg, NC)
528th Combat Stress Control Detachment (Ft Bragg, NC)

Providing care to Somali nationals (refugees) or to relief workers operating within the theater was not part of this medical mission.[3] Neither was getting involved in the actual humanitarian relief activities within the various sectors—that was to be the purview of the NGOs.[4] In theory, all the U.S. military's medical resources were to go in support of U.S. forces during this operation.

In addition, unlike in the Balkans where there was a close working relationship with the local hospitals in Zagreb, which were quite capable by Western standards and thus able to support Echelon IV care, in Somalia the local medical infrastructure had been completely destroyed, evacuation times were longer than doctrine normally calls for, and there were long distances between the theater Echelon III hospital in Mogadishu and fixed facilities in neighboring countries.[5]

What Happened to the Medical Mission and Medical Support During the Operation

Figure 4.1 illustrates how the medical mission and medical structure evolved over the course of this operation. Shown are the total number of PROFIS medical personnel (N = 991) in-theater over the 17 months of this deployment.[6] The x axis represents the number of months in-theater beginning with November 1992 and ending with

---

[3]The Somali nationals received the bulk of their care from the Swedish hospital. The Swedes provided two levels of care: one for U.S. troops and UN forces, and another for Somali nationals that was more commensurate with the level of care within that country.

[4]At the height of the relief effort, Somalia had over 60 NGOs or humanitarian relief organizations (HROs) operating within the eight sectors of the country.

[5]Specifically, it was a fourteen-hour flight from Mombasa, Kenya, to the Army hospital in Lanstuhl, Germany. In Mombasa, an international joint task force had an air transportable clinic (ATC) set up to provide support to the U.S. and coalition troops in Somalia requiring evacuation out of the theater. This clinic only had outpatient capabilities but had also contracted with the local hospital. The Mombasa operation ended around mid-March 1994 for all intents and purposes and remained closed for three months. Following the October 1994 Ranger incident, the Mombasa operation reopened to support the U.S. and UN forces in Somalia. Source: Interview with LTC Courtney Scott, M.D., JTF Surgeon for OPR, serving in the U.S. Air Force.

[6]The figure shows officers only. Although there were some enlisted PROFIS personnel in-theater, they were relatively few and not typical of the kinds of PROFIS personnel who usually deploy. Enlisted personnel are excluded from Figure 4.2 as well.

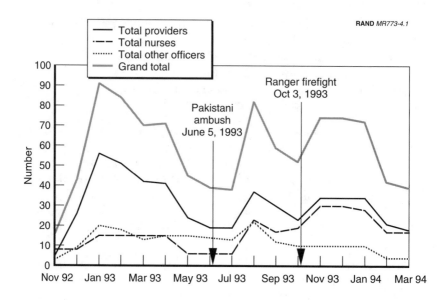

Figure 4.1—How the Medical Mission and Support Changed
During the Operation

March 1994. The y axis represents the number of PROFIS personnel
in-theater at the midpoint of each month.[7]

Initial deployment: November 1992 to January 1993. Although the
first rotation began officially in January 1993 as ORH got under way,
there was a preliminary deployment of medical personnel beginning
in November 1992 (Figure 4.1). The 10th Mountain Division's organic medical assets were among the first Army medical personnel to
arrive in-theater on 18 December 1992.[8]

---

[7]The provider category includes surgical, primary care and medicine, mental health,
and dental specialties. The "other officers" category includes administrative and
health services specialties, preventive medicine, and ancillary support specialties (e.g.,
pharmacy, laboratory science, infectious disease, diagnostic radiology, dietetics,
optometry, and pulmonology). The nursing category includes operating room, medi-
cal-surgical, and clinical nurses.

[8]The division's organic medical assets were supported by a Navy Corps Collecting and
Clearing Company and by the USS *Tripoli.* An Air Force air transportable hospital
(ATH) had also been established in Cairo to serve as the intermediate link in the

Health service support initially was to be the responsibility of the Commander, Joint Task Force-Somalia (CJTF-S) Surgeon's Office, backed up by the Navy and Marines (MARFOR). U.S. forces were to receive Echelons I and II care from organic medical assets, with the USS *Tripoli* serving as Echelon III support.

The rapid rise in Army medical personnel seen during December 1992 and January 1993 represents the transitioning into the theater of various Army medical elements in anticipation of the Army assuming theaterwide health service support.[9] These elements included an advance assessment party, headquarters of the 62nd Medical Group, and the 86th EVAC, which arrived in-theater on 24 December 1992. Headquarters, 62nd Medical Group arrived on 29 December 1992 to initiate plans and activities to assume responsibility for theaterwide medical support. The 86th EVAC, a 104-bed hospital, took the first rotation into Somalia, arriving in-theater on 18 January 1993.[10] With the arrival of Army corps-level assets, the Army assumed Echelon III medical care and hospitalization for U.S. troops.[11]

Starting with the peak of 91 PROFIS personnel in January 1993, medical support in the first rotation plateaued at around 70 personnel by March 1993, which probably represented the true level of support during most of ORH. The peak in medical support corresponded to the peak in the number of U.S. troops that occurred on 16 January 1993.

First rotation: January 1993 to May 1993. During the first rotation, the overall mission was primarily a humanitarian relief effort, with

---

strategic aeromedical evacuation chain. Source: "Operation Restore Hope Medical After Action Report," DASH-HCO-P, MAJ Michael Gunn.

[9]In addition, several PROFIS physicians were initially sent over to augment the 10th Mountain Division's organic medical assets.

[10]Three types of hospitals were deployed to Somalia: an evacuation (EVAC) hospital, a field hospital (FH), and a combat support hospital (CSH). The EVAC is designed to provide hospitalization to all classes of patients in the combat zone. At its maximum capacity, an EVAC consists of four intensive-care wards, eight intermediate care wards, and ten minimal-care wards. Note that the nominal bed capacity of a hospital (104 beds in this case) can be misleading, since not all of its wards may be set up.

[11]Theaterwide medical support responsibility was transferred from the CJTF-S Surgeon's Office to the Army Commander, 62nd Medical Group on 28 January 1993.

the medical mission quickly evolving into one of primarily providing routine care to U.S. troops. Although U.S. troops and the 86th EVAC itself took a fair amount of sniper fire during this period, Somalia still represented a relatively benign environment with few combat casualties.

In addition, during this period the Army did take care of some civilian casualties. These patients often were brought in by enlisted troops rather than by the medical staff.[12] However, it rapidly became clear that the NGOs were not happy with the Army providing medical care to the Somalis, fearing the loss of some of their own "business."[13]

During the first rotation, the demand for medical services was lower than initially anticipated. The outpatient rate, for example, was roughly half that of the predicted rate, and the DNBI rate was also lower than expected. By the end of this rotation it appeared that the Army did not require nearly as many medical assets in-theater as what had initially been brought in.

Second rotation: May 1993 to August 1993. On 1 May 1993, the second rotation and the transition from ORH to OCH occurred, with the 42nd Field Hospital (FH), a 32-bed facility, taking over from the 86th EVAC.[14] In early May, the 42nd Medical Task Force (MTF 42) assumed JTF medical responsibilities as the senior medical headquarters. As shown in Table 4.1, MTF 42 was to provide command and control, Echelon III care and hospitalization (42nd Field Hospital), medical logistics (147th Medical Logistics Battalion), outpatient ser-

---

[12]The policy was for the U.S. forces to treat whomever they injured (e.g., if an Army truck ran over a Somali citizen). There was some fratricide by Somalis; individuals would push relatives underneath the wheels of a Army vehicle in order to obtain monetary compensation from the Army for an injury or death.

[13]Based on the presentation of the CENTCOM Surgeon, Uniformed Services University of the Health Sciences (USUHS) conference, November 1994, Washington, D.C.

[14]A field hospital is designed to provide hospitalization for patients within the theater of operations who require further stabilization prior to evacuation and convalescent care to patients who will be returned to duty with their field unit. It can accommodate up to 504 patients. At its maximum capacity, a field hospital includes two ICU wards, seven intermediate nursing care wards, one ward for neuropsychiatric care, two minimal-care wards, and seven patient support sections providing convalescent care. A field hospital is typically staffed with 19 Medical Corps officers and 57 Nurse Corps officers.

vices and ground evacuation (61st Area Support Medical Battalion), helicopter aeromedical evacuation (45th Medical Company—Air Ambulance), preventive medicine (105th Medical Detachment—Sanitation), veterinary support (248th Veterinary Detachment), and mental health and combat stress control (528th Combat Stress Control Detachment).

During the second rotation, we observe a tailoring back of the medical personnel such that by May 1993, the total number of PROFIS medical personnel in-theater had been greatly reduced from the high of 91 to about 45. There are a number of reasons for this, but the primary one is that medical support tracks the decline in U.S. troops to support. The number of U.S. troops in-theater peaked at 25,400 on 16 January 1993 and then steadily dropped to about 10,000 by late April 1993. After the first rotation, the number of U.S. troops in-theater averaged around 4,000 and 4,200 during the second and third rotations that took place during OCH. Further, we see a narrowing in the difference between the number of providers and nurses in-theater at this time.[15]

When the operation shifted from ORH to OCH, the United States continued its medical mission of supporting U.S. forces involved in this humanitarian and peace enforcement effort and of providing emergency-only support to other UN forces within the theater. Although this was a humanitarian action, throughout ORH and OCH the mission was characterized by constant sniper fire, punctuated by periods of combat.

However, one event changed the overall mission and the medical support mission. On 5 June 1993, 24 Pakistani soldiers were ambushed while taking part in the UN peacekeeping operations within Somalia. While responsibility was never proven, the ambush was believed to have been directed by General Mohamed Farah Aideed, one of the dominant clan leaders in the country. Following the Pakistani ambush, tensions within the theater rapidly escalated, with the mission going from one of "guarding the beach and handing out

---

[15]However, the dropoff in the number of personnel during the second rotation from May to mid-August 1993 indicates that the 42nd FH did not deploy with its full complement of medical personnel.

food" (i.e., humanitarian relief) to one of being in a combat-like environment, now with a tangible threat.[16]

Although during May the amount of sniper fire into the embassy compound where the 42nd Field Hospital was located had been steadily increasing, after the Pakistani incident U.S. troops were increasingly locked down; with the Italian, French, and Norwegian medical personnel coming under fire, concern grew for U.S. medical personnel. As a result, for security reasons, the 42nd Field Hospital's staff were closely tied to the embassy compound. The hospital began seeing an increasing number of U.S. casualties, and there was a substantial increase in the level of stress among U.S. troops. Also, following the Pakistani ambush, there was a growing resentment among the medical staff at having to treat Somali patients. As a result, the hospital commander tightened up on the treatment of civilian casualties.

Third rotation: August 1993 to March 1994. On August 15th, the third rotation began with the 46th Medical Task Force (MTF 46) assuming JTF medical responsibilities for the third and longest rotation. MTF 46 comprised 270 PROFIS personnel and was tasked to provide health services and hospitalization to the U.S. contingent and to UN forces on an emergency basis, as well as treatment to Somali nationals who were wounded as a direct result of confrontation with UN forces. In addition, the 10th Mountain Division during the third rotation had health service support to assigned division personnel.[17] At this point, the 46th Combat Support Hospital (CSH) took over hospitalization care from the 42nd FH.[18]

In Figure 4.1, the steep rise in PROFIS personnel during late July and the decline in September represent the transition of medical units

---

[16]Patric J. Sloyan, "The Somalia Endgame—How the Warlord Outwitted Clinton's Spooks," *Washington Post*, 3 April 1994.

[17]MAJ Michael Gunn, *U.S. Army Medical Department Operations in Somalia—Update*, 15 September 1993.

[18]A combat support hospital is the most comprehensive hospital unit within a theater, being designed to provide hospitalization for up to 296 patients within the combat zone. At its maximum capacity, a combat support hospital comprises eight ICU wards, seven intermediate nursing care wards, one ward for neuropsychiatric care, and two minimal-care wards. A combat support hospital at its maximum capacity is staffed with 33 Medical Corps officers and 120 Nurse Corps officers.

into and out of the theater. This transitioning of the medical support took somewhat longer than the previous one. This may have been due partly to the fact that PROFIS personnel for the 46th CSH were pulled from a number of different MEDDACs and MEDCENs across CONUS and so did not arrive in-theater as a single group. Medical personnel had to wait until their replacement had been identified and then sent over before being able to rotate out of the theater.[19] As a result of this rotation policy, divisions began to develop between the newer and older staff in the theater. For example, by the time of the third rotation, many who had been in-theater longer and had experienced the escalation of tensions felt a lot of anger with the Somalis and were reluctant to treat Somali patients—the growing resentment discussed above. The newer hospital staff, who had not yet experienced any of these incidents, did not understand the anger. There was one report that during the third rotation, the older staff considered segregating the Somali patients into a separate ward and having only the newer staff provide care to them.[20]

During the transition between the second and third rotations, we also see for the first time an increase in the absolute numbers of nursing staff in-theater, while the total number of providers declines to its lowest point (see Figure 4.1). By the beginning of October, the number of providers and number of nursing staff in-theater were roughly equal (23 and 19, respectively), with the total number of PROFIS personnel between 55 and 60 individuals.[21]

In addition, the medical mission during this period continued to be primarily one of providing routine care. The outpatient visit rate had remained at half its predicted rate. Further, because of increased tensions in the theater, the hospital staff were mostly confined to the embassy compound for security reasons, with little opportunity to undertake sector visits or volunteer with the HROs.

---

[19]During this period, U.S. troops moved a great deal in and out of theater.

[20]The third rotation was also plagued by other morale problems that were related either to the rotation policy or to the fact that the 46th CSH was scheduled to deactivate upon its return to OCONUS.

[21]The composition of the CSH was clearly tailored, since the TOE requirement would normally call for 33 Medical Corps officers and 120 Nurse Corps officers, a 1:3 ratio for these hospitals.

During August and September 1993, tensions in the theater continued to rise, with an increase in the number of demonstrations and
displays of weaponry. The 46th CSH took more and more rounds
into the embassy compound, and Somalis were attacking UN personnel, the media, and relief workers. In September, the 362nd En-
gineering Group was ambushed, and on 1 October 1993, a U.S. helicopter was shot down, killing three U.S. soldiers.

The Ranger firefight on 3 October 1993 marked another key change
in the overall mission. During the firefight with supporters of General Aideed, 18 American Rangers were killed and 77 wounded. The
firefight was a culmination of an extended manhunt by U.S. troops to
capture General Aideed for his alleged role in masterminding the
June 5th ambush of 24 Pakistani peacekeepers. In addition to U.S.
casualties, an estimated 300 of Aideed's followers were killed and
another 700 wounded in this firefight. Because of the high number of
American casualties incurred in this incident, U.S. public opinion
turned strongly against a continued U.S. presence in Somalia.[22]

As noted above, by this point the number of providers in-theater was
at its lowest level (Figure 4.1). As a result of the Ranger firefight, the
Army again increased the number of beds and medical personnel in-
theater and would sustain this level for the remainder of the operation.

After 4 October 1993, the entire theater shut down and security measures increased, including greater restrictions on the movement of
AMEDD personnel within the theater. Tensions in-theater remained
high throughout the remainder of the operation. Starting in January
1994, we see the reduction in medical support starting to occur as the
March 1994 deadline for withdrawal approached (Figure 4.1).

When we look across the 17-month deployment, we can summarize
the change in the medical mission as follows: The initial mission began as a humanitarian relief effort in a relatively combat-free zone, in
which the primary medical mission was to support U.S. troops and
possibly provide care as well to Somalis, coalition forces, and NGOs;
it evolved to a mission of supporting U.S. forces only and UN troops
on an emergency basis in an increasingly combat-like environment.

---

[22]George J. Church, "Anatomy of a Disaster," *Time*, 18 October 1993.

Although the medical need much of the time was mostly for primary care, there was the potential, especially in the latter half of the operation, for combat casualties.

What Happened to the Medical Structure During the Operation

When we examine the medical support during the operation—especially between the first two rotations—we see that Army had more medical assets in-theater than it needed. The fact that the medical support was at its highest level in-theater during the first rotation was partly the result of the need to support a larger number of U.S. troops at the beginning. Initially, the number of U.S. troops requiring medical support was high; it peaked at approximately 25,400 troops on 16 January 1993, and then steadily dropped to about 10,000 by late April 1993. After the initial rotation, though, the number of U.S. troops in-theater averaged between 4,000 and 4,200 during OCH (UNOSOM II).[23]

However, there was another reason for the excess medical capacity brought in initially. Although humanitarian medical support to Somalis and to NGOs was not specified as a task in the initial medical mission statement, these objectives were taken into consideration in tailoring the medical support for this operation. More specifically, this expectation probably influenced the large number of providers sent and the mix of specialties brought in. It may also explain why there was a much greater number of providers in-theater relative to nurses during ORH: the range of specialties was increased to accommodate the expected need of these other patient populations. The large proportion of providers relative to nurses was clearly an intentional decision, since an EVAC hospital—like the deployed 86th during the first rotation—typically has a ratio of one Medical Corps officer to every three Nurse Corps officers. In addition, the large number of providers in the initial phase may have been a function of the deployment of advance assessment teams.

---

[23]The number of troops includes all three services. LCDR Gradisher, DoD Public Affairs Office, was the source on troop end strength for UNOSOM I and II.

Although the number of beds and personnel was subsequently cut back, the structure tended to follow the key events. Following the June 1993 Pakistani ambush during the second rotation, the Army increased the number of beds and medical personnel in-theater.

The number of beds and personnel tapered off once again over time, so that by the time of the Ranger firefight in October 1993 during the third rotation, the number of providers in-theater was at one of its lowest levels. In response, the Army again increased the number of beds and medical personnel in-theater and sustained this level through the beginning of 1994.

See the "requirements" section below for a more complete discussion of the evolution of the number and mix of providers during the operation.

DEMAND FOR SERVICES

Expectations as to Patient Demand at the Outset

In planning the medical support for this humanitarian operation, the requirement was for a corps-level package to support a brigade to assault an airhead with an unknown level of combat.[24] That is, although the medical planners were aware that this mission was primarily humanitarian in nature, given the lawlessness of the situation and the number of warring factions, it was difficult to assess to what degree U.S. troops might encounter armed opposition in accomplishing this mission.

Therefore, in terms of patient demand the expectation was a requirement for routine primary care but also a requirement for an unknown level of trauma care. Given that the planners did not know what level of combat might be associated with this mission and the large number of U.S. troops to support during ORH, it was difficult to estimate the demand for trauma care and surgical services during the first phase of the Somalia mission.

---

[24]Interview with COL David Nolan, January 1996.

Patient Demand During the Operation

Table 4.2 shows the total number of admissions and outpatient visits across the rotations. We grouped outpatient clinic visits and emergency room (ER) visits into a single category, since the coding of these two types of visits did not appear to be consistent across the three rotations.

Table 4.2 indicates that the peak in admissions and outpatient visits occurred during the first rotation, between January 1993 and May 1993, when the largest number of U.S. troops were in-theater. Still, as noted earlier, the demand for medical services during this rotation was lower than initially anticipated. After the first rotation, although not shown, the patient load became fairly constant over the remainder of the deployment, with the number of admissions averaging 166 per month and the number of outpatient visits averaging 1,000 per month.

We also examined the breakdown of the admissions shown in Table 4.2 by clinical service to understand differences in the proportion of admissions across services for each of the three rotations. To do so, we grouped the inpatient admissions into four categories: internal medicine, surgery, ob/gyn, and psychiatry. Table 4.3 shows the following trends over the course of the deployment in terms of the relative distribution of admissions across clinical services: The proportion of internal medicine admissions decreased over time from a high of 62 percent in the initial months to around 40 percent by the

Table 4.2

Comparison of Total Number of Admissions and Outpatient Visits by Rotation for ORH and OCH

Rotation	Number of Outpatient Visits	Number of Admissions
1st rotation—86th EVAC	4,914	971
2nd rotation—42nd FH	2,906	361
3rd rotation—46th CSH	4,903	568

NOTE: The outpatient visits category refers only to the 86th EVAC, the 42nd FH, and the 46th CSH (i.e., the data exclude sick call for the outlying field units).

Table 4.3

Percent of Total Admissions by Clinical Service for Rotations in ORH and OCH

	1st Rotation: 86th EVAC			2nd Rotation: 42nd FH			3rd Rotation: 46th CSH				
	1/93	2/93	3/93	5/93	6/93	7/93	9/93	10/93	11/93	12/93	1/94
Internal Medicine	62%	61%	43%	71%	45%	48%	38%	24%	38%	35%	42%
Surgery	35	35	52	25	48	44	50	69	51	61	51
Ob/Gyn	3	4	3	2	5	4	9	2	9	3	4
Psychiatry	1	1	2	2	2	5	4	4	2	1	4
Total	333	350	288	167	110	84	101	186	125	99	57

NOTE: The patient data cover the period between January 1993 and January 1994 and do not include the first few months (November and December 1993) or the latter few months (February and March 1994). Also, patient data were unavailable for April 1993 and August 1993, the two months when the rotation of U.S. troops and hospital units into the theater took place.

third rotation. In contrast, the proportion of surgical admissions steadily increased over time. As a result, the relative distribution of admissions between internal medicine and surgery changed over time from 62 percent internal medicine/35 percent surgery at the beginning of the first rotation to 42 percent internal medicine/51 percent surgery by the end of the third rotation. Part of the increase in the proportion of surgical admissions may have been related to an increase in the proportion of Somali patients being treated by the AMEDD, since the physicians were selectively treating foreign nationals who required surgery. (See the next section for a discussion of populations treated.) However, the fact that tensions within the theater were increasing over time and, thus, the probability of combat casualties likely explains much of the rise over time in the proportion of surgical admissions.

Some of the small proportion of ob/gyn admissions represented female soldiers who had tested positive for pregnancy. Because the antimalarial drug, mefloquine, could not be administered to these soldiers, they were admitted and put under netting to protect them until they could be rotated out of the theater.[25]

Expectations of Populations Served at the Outset

As mentioned earlier in the discussion of the mission statement, the U.S. military was expecting to serve primarily U.S. forces during this operation. It was also expecting to serve coalition forces on an emergency-only basis. Providing care to Somali civilians, refugees, or relief workers operating within the theater was not part of this medical mission. Neither was getting involved in the actual humanitarian relief activities within the various sectors.

Populations Served During the Operation

To examine the populations served during the operation, we broke the above admission data down further by the following categories: U.S. personnel, foreign military, and foreign civilians. As shown in Table 4.4, the peak in the total number of U.S. personnel admissions

---

[25]Interview with a nurse from the 46th CSH, March 1994.

Table 4.4

Number and Percent of Total Admissions by Populations Served
for Rotations in ORH and OCH

	1st Rotation: 86th EVAC			2nd Rotation: 42nd FH				3rd Rotation: 46th CSH			
	1/93	2/93	3/93	5/93	6/93	7/93	9/93	10/93	11/93	12/93	1/94
U.S. personnel	308 92%	308 88%	247 86%	130 77%	72 65%	64 74%	85 83%	166 89%	92 73%	58 61%	42 72%
Foreign military	2 1%	9 3%	16 6%	22 13%	34 31%	16 21%	4 4%	5 3%	5 4%	13 13%	5 9%
Foreign civilians	23 7%	33 9%	25 9%	15 9%	4 4%	4 5%	12 13%	15 8%	28 23%	28 29%	10 19%
Total	333	350	288	167	110	84	101	186	125	99	57

NOTE: Patient data were not available for April and August 1993, when the rotation of the hospitals occurred.

occurred in January and February 1993. The table also shows the fluctuations in the admission of U.S. personnel over time, with the jump in admissions to 166 in October 1993 corresponding to the Ranger firefight. Table 4.4 also indicates that the number of admissions of foreign personnel (military or civilians) was relatively low over the course of the Somalia deployment. The increase in the number of foreign military admissions in June 1993 corresponds to about the time of the Pakistani ambush.

When we look at the proportion of total admissions that were U.S. personnel, foreign military, or foreign civilians, we see that the proportion of total admissions that were U.S. personnel started at just above 90 percent and then declined somewhat over time. Table 4.4 also indicates a seesaw nature in the admission pattern of U.S. personnel.[26]

In comparison, the proportion of total admissions that were foreign military was about 10 percent for most of the deployment. The increase to 30 percent in June 1993 corresponds to the time of the Pakistani incident. The fall in the proportion of admissions of U.S. personnel during this same period is coincidental, being more related to the rotation of U.S. forces at this time and a decline in the absolute numbers of U.S. personnel in the theater between the first and second rotations.[27]

The proportion of total admissions that were foreign civilians went from 8 percent initially to a high of 29 percent in December 1993. Thus, the AMEDD did treat Somali nationals throughout this deployment, with the proportion of Somali inpatients steadily increasing over time.

We also compared differences in average length of stay (LOS) between U.S. personnel, foreign military, and foreign civilians for each rotation. Figure 4.2 shows the average LOS for each rotation by hospital and by patient category. Overall, the average LOS during this

---

[26]Grouping U.S. military and U.S. civilian personnel together in one category did not affect any of the results reported herein. One could argue that U.S. civilian contract and federal employee personnel would tend to be older, more likely to have chronic conditions, and more likely to be in poorer health than U.S. military personnel. However, their numbers were few and so did not affect any of our results.

[27]Recall that the number of U.S. troops in-theater went from a high of 26,400 in January to 4,200 by June 1993.

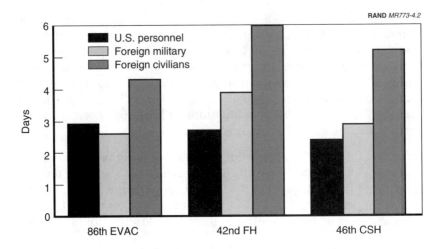

Figure 4.2—Average Length of Stay by Rotation of
Patient Populations Served

deployment by patient group was 2.7 days for U.S. personnel, 3.4 days for foreign military, and 4.9 days for foreign civilians. Although the overall average LOS for each of the rotations was 3 days, when we broke out the LOS data by patient categories, we found that foreign military and foreign civilians tended to have longer lengths of stay on average than U.S. personnel for each of the three rotations.

Although the numbers are small for foreign military and foreign civilians and thus any trends identified need to be interpreted with caution, the results shown in Figure 4.2 suggest that foreign military and foreign civilian patients tended to be more severely ill and more resource intensive than U.S. personnel.[28] In the case of the Somalis, these patients tended to be more severely ill, partly because the AMEDD medical staff at times deliberately selected the sicker patients to treat. The longer LOS for foreign military personnel also reflects the fact that it was difficult for the AMEDD at times to repatri-

---

[28]One of the nurses interviewed from the 46th CSH corroborated this suggestion. She made the observation that at one point, the Somali nationals constituted a relatively small percentage of the total number of inpatients, yet they required 80 percent of the nurses' time, since they were also the sickest patients in the hospital.

ate injured coalition soldiers.[29] As a result, the AMEDD ended up hanging onto coalition patients longer than would normally be expected.

Overall, the data suggest that although the foreign patients (military and civilians) made up a relatively low percentage of the total admissions, they used a disproportionately larger percentage of a hospital's resources. For example, many of the coalition forces in Somalia did not maintain tight controls over their food and water supplies or enforce good sanitation within their living quarters. As a result, some troops had a lot of problems with diarrhea and upper respiratory infections. In Somalia, the high malaria rate among some troops (e.g., Pakistanis) was the result of a number of factors: (1) no preventive medicine precautions were undertaken before or during the deployment by some contingents; (2) in some cases, inappropriate chemoprophylaxis was being used (i.e., outdated or less effective medications); (3) some countries were unable to afford the more expensive, modern antimalarial drugs; and (4) the United States had bought up much of the world's supply stocks of mefloquine, making it difficult for other coalition partners to obtain this drug even if they could afford it.

In the last chapter we discussed these issues at length, using patient data from UNPROFOR. In Croatia, the AMEDD's mission was to provide hospitalization care (Echelon III) to all the coalition forces, so the number of foreign military patients is larger in that operation and better illustrates the differences in resource intensity between U.S. personnel and foreign patients.

REQUIREMENTS TO MEET THE DEMAND FOR SERVICES

To understand how well the requirements matched the demand for services, we examined how the specialty mix of PROFIS personnel changed overall and then within categories. Specifically, we compared the percentage of total PROFIS personnel who were providers, nurses, or other officers. We then broke down the provider and

---

[29]Reasons for the problems in repatriating coalition soldiers included the country of origin lacking MEDEVAC capabilities, among others. The problem of repatriation is discussed more fully in Chapter Three.

nursing categories into their various components to look at changes in specialty mix within categories over time. Although some of the changes in specialty mix seen are likely a function of the type of hospital deployed, not all changes may be attributable to the TOE requirement alone.

How the Overall Mix Between Providers and Nursing Staff Changed

In January 1993, at the official beginning of ORH, and throughout the first rotation, approximately 60 percent of the total PROFIS personnel in-theater during the first rotation were providers (see Figure 4.3). This proportion then leveled off at around 50 percent during the second and third rotations.[30]

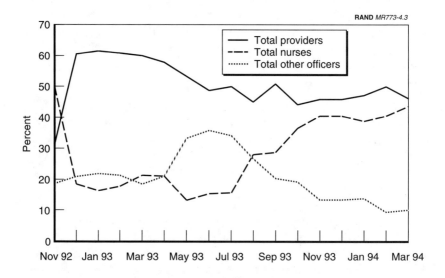

Figure 4.3—How the Proportion of Medical Personnel Changed Overall During the Operation

---

[30]As noted earlier, the high proportion of providers during the initial rotation may have been due to the expectation of being tasked to provide medical care to civilians and humanitarian relief volunteers as well as U.S. troops.

In contrast, the proportion of nursing staff started out low, at around 17 percent in the first half of the operation, and then increased over the first rotation, eventually leveling off at about 45 percent in November 1993. As a result, by the third rotation there was approximately a 1:1 ratio of providers to nurses.[31]

The increase in the proportion of nurses over time further supports the evidence shown above on patient demand that the medical support evolved into a primary care mode over time. Figure 4.3 also indicates that the proportion of administrative and other officers declined over this same period.

How the Specialty Mix of Providers Changed Over Time

Figure 4.4 shows how the mix of provider specialties changed over the course of the Somalia deployment. The most interesting story is the seesaw effect seen in the mix between primary care physicians and surgeons. At the beginning of the first rotation, primary care physicians clearly dominated the mix over surgeons, 45 percent versus 29 percent, respectively. Still, as the figure shows, this relationship was already trending in opposite directions, since in December 1992, the primary care/surgeon mix was 50 percent versus 19 percent. In fact, there was a steady increase in the proportion of providers who were surgeons during the first rotation, such that by the second rotation (months 7–9), the proportions of surgical and primary care providers were approximately equal.

In August 1993, with the transition in of the 46th CSH, the specialty mix changed once again, with a decline in the proportion of providers who were surgeons. As a result, by the time of the Ranger firefight in October 1993, only 26 percent of the total number of providers in-theater were surgeons. In contrast, 61 percent of providers were primary care.

As shown in Figure 4.4, the 61 percent in October 1993 of the deployment represents the peak for primary care; for the remainder of

---

[31]This mix (high proportion of providers/low proportion of nurses) may have been related in part to the first two hospitals deployed being an EVAC and a Field Hospital.

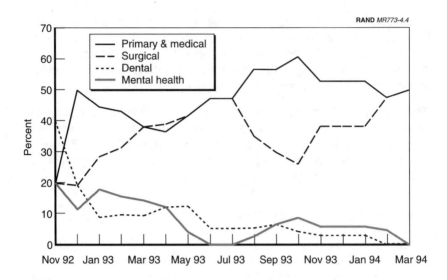

Figure 4.4—How the Mix of Providers Changed During the Operation

the rotation, the proportion of surgeons began to grow again, such that by month 16, the two categories of providers were equal.

Still, the increase in the proportion of primary care providers over time (combined with the increase in the proportion of nurses) supports our hypothesis that the medical mission had evolved into a primary care mode by the third rotation.[32] There is also a mismatch between demand and supply here. Recall that Table 4.3 shows that the demand for surgery grew larger during the third rotation, rising to 69 percent of the admissions during October 1993, when the Ranger incident occurred. However, as Figure 4.4 shows, in October 1993 only 26 percent of the providers were surgeons. While there is a steady increase in the proportion of surgeons after October 1993, no doubt in response to the increased demand, the proportion of surgeons steadily declined from July to October 1993 (from 47 to 26 percent) as the mission evolved toward primary care, despite the

---

[32]Recall that during the initial rotation it was clear that the medical need was fairly low, with few combat casualties.

earlier growing demand for surgery from July to October 1993 (from 44 to 69 percent).[33]

Thus, we see that changes in the medical need in Somalia did not always closely match changes in the mix of providers within the theater of operation, particularly as the threat level increased.

Figure 4.4 also shows that the proportion of PROFIS personnel who were mental health providers was initially too high and then went too low.[34] In the first rotation, the mental health providers started at approximately 18 percent of the total number of PROFIS providers in-theater in January 1993, declined to around 12 percent by April 1993, and then declined further during the second rotation (to 0 percent in June and July 1993), before rising to about 5–6 percent for most of the third rotation.[35]

Recall that during the first rotation, the theater was still relatively benign, and tensions had not yet started to heat up. In the second rotation, the combat stress assets were subsequently tailored back. As a result, however, the mental health assets in-theater from this point onward were probably too low; after the Pakistani incident,

---

[33]Although it is not shown in the figure, when we broke the surgical specialties down further into their various components (i.e., ob/gyn, general/thoracic/ orthopedic, and other surgical specialties), we found that the mix of surgical specialties in-theater remained fairly constant over the deployment, even though the percentage of providers who were surgeons initially increased, leveled off, and then decreased over time. Of the 184 surgeons in-theater over the 17 months of this deployment, approximately 45 percent were general/thoracic/orthopedic, another 45 percent fell into the "other" category, and less than 10 percent were ob/gyn.

[34]This was due partly to ORH representing the first deployment of the combat stress teams, before the permanent assignment of personnel to the 528th Combat Support Company (CSC) could be completed. The 528th CSC has noted that a key lesson learned from the Somalia deployment was the need to send a smaller rapid deployment assessment team over first to assess the mental health threat and to tailor the combat stress support.

[35]For the first rotation, a twelve-person prevention section was deployed to Somalia, so that the mental health team initially comprised thirteen officers and one enlisted personnel. For the second rotation, this number dropped to one officer and one enlisted personnel, too few given the increase in violence within the theater during this period, per CPT Eric Cipriano. The 528th Medical Detachment (Combat Stress) noted in its after-action report that a minimal mental health team of two officers (one prevention and the other restoration) and two enlisted personnel was required during the second rotation, as was the flexibility to augment the staff as the threat level increased.

conditions in the theater became more combat-like, and the concomitant level of stress also increased among U.S. troops and AMEDD personnel. The stress level would remain high for the duration of the deployment.[36]

As shown in Figure 4.4, the dental assets grew from 9 percent to around 12 percent during the first rotation, dropping off to around 5–6 percent during the second rotation, and ultimately falling further to around 3 percent during the third rotation. The dentists were some of the busiest of all providers, with one individual estimating that as much as 20 percent of all outpatient visits during the first rotation were for dental care.[37] Coalition forces in particular sought dental care from the Army.

When we consider the match between demand and requirements, was the medical support provided too much, too little, or just right for Somalia? The answer depends on one's perspective. The medical support was more than adequate for most periods when not much was happening in-theater.[38] In the case of the mass-casualty incident (i.e., the Ranger firefight of October 1993), however, the AMEDD was stretched thin. Within the first 34 hours, the 46th CSH received 36 cases. Most of the cases seen were trauma patients (with limbs blown off, bad burns, etc.).[39] There was a total of 110 casualties during this period, all requiring surgery. Recall, as discussed above, that this incident occurred at one of the times when the United States had the fewest number of surgeons in-theater and the fewest number of providers.[40] At this point in the operation, the 46th CSH only had assigned two general surgeons and one orthopedic surgeon. As a result, the AMEDD ended up transferring four orthopedic cases

---

[36]Based on interviews with the 528th Combat Stress Control Detachment, Fort Bragg, NC.

[37]Interview with MAJ Winton Carter, 257th Dental Detachment, first rotation into Somalia. As in Somalia, the dentists in Zagreb, Croatia for the UNPROFOR mission were in high demand by the coalition forces.

[38]One exception, as shown above, was the level of combat stress support in the theater for the latter half of the deployment.

[39]Casualty numbers are based on the interview with MAJ John Holcomb, M.D.

[40]One of the surgeons was a Special Operations Forces surgeon sent over specifically to provide support to the Rangers.

to the Swedish hospital, which was co-located with the Army hospital in the embassy compound.[41]

OVERALL CASE-SPECIFIC OBSERVATIONS

When we look across the experience of the medical mission in Somalia, a number of observations emerge. On the whole, these center around the problems of planning medical support given the variable demand, problems with treating civilian populations, and issues of training.

Planning the Medical Support for Variable Demand

Somalia was an example of an OOTW characterized by low-intensity conflict but also having the potential for combat. As a result, Somalia illustrated how difficult it is to plan the medical support for an operation in which patient demand is characterized by "peaks and valleys" and how key events may often drive the medical support and planning process once in-theater. In such a situation, commanders need to be able to react to events as they unfold. At the same time, a commander must be able to make the best use of his medical assets. As shown in the case study, the medical need may not always closely match the type and amount of medical support in-theater. In these operations, then, flexibility in planning becomes key.

The initial planning process for a deployment and the decision on what mix of specialties may be required typically takes into account five key factors: (1) the nature and level of the medical threat; (2) the medical mission statement and number of troops to support; (3) the doctrinal employment of units and the organizational capability of those units; (4) what augmentation of the TOE requirement for a hospital may be required; and (5) which unit is ready to be deployed and best meets the requirements of the specific mission.[42]

---

[41]Because the Americans and the Swedes had set up weekly case exchange meetings to facilitate standards of care in-theater and encourage a professional dialogue, the 46th CSH knew well in advance that the Swedes would be able to handle these cases.

[42]The decision about which unit to deploy is also a subjective one and often involves the input of a number of different entities. For example, for Somalia the planning sequence included the command elements of CENTCOM, FORSCOM, the 3rd Army,

This yields a certain level of hospital with certain capabilities. If there is some excess capacity as a result of this process, then one may choose to provide some additional level of support to other coalition forces or civilians, or one may want to tailor back. For Somalia, after the first rotation, the equipment and supplies were larger in place than was needed and the mix of medical personnel was too heavy on the surgical side. The tailoring back of the support that occurred at the beginning of the second and third rotations was based on operational needs rather than on organizational structure.[43]

During an operation, the prerogative of the theater commander is paramount. The in-theater medical commander needs to be given the flexibility to tailor or modify the medical support brought in as he sees fit to meet the needs of the day-to-day operation. However, he also needs high-level guidance to ensure that a minimum level of medical support is maintained in-theater throughout the deployment. The best way to accomplish this may be through doctrine.

Given the demand variability associated with some OOTW, the key question from a planning perspective is how to determine the right level of medical support required and the right mix of medical personnel. The lesson from Somalia for DoD and Army planners may very well be to staff for a little more than the average and then extend for the surge. In OOTW, one cannot staff for the worst-case scenario and support it. However, by staffing for a little more than the average, one is ensured of an extended capability in-theater.[44] The idea would be to have enough medical personnel in-theater to get through the initial 24–48 hours of a mass-casualty situation. One would not want to staff just for the average, because this would take the elasticity out of the system.

It is also clear that the size of hospitals in OOTW is small, which means they may easily be overwhelmed in a mass-casualty situation.

18th Airborne Corps, and the 44th Medical Brigade. The 44th Medical Brigade developed the operation plan (OPLAN) for ORH and then briefed the 62nd Medical Group, which developed the actual execution plan.

[43]Interview with COL David Nolan, former executive officer, 44th Medical Brigade, January 1996.

[44]However, in terms of medical supplies and equipment, one wants to plan for the worst-case scenario so there will be in-theater what is needed in case of a mass-casualty situation.

This suggests that medical evacuation becomes a top priority in these operations. The Army may want to consider having on call a backup team of medical personnel based in Italy or Germany or located on a carrier that can be flown into the theater within 24 hours to support the medical personnel already in-theater if there is a mass-casualty situation. The airplane transporting medical personnel could then be used to evacuate patients.

In addition, the Army might want to consider developing a staging team along the Air Force model, in which 15–20 medical personnel and a 25- to 30-bed tent capability are equipped and trained to take casualties and move them out. Such a team would be capable of accepting casualties for initial treatment and then evacuating them out of the theater. Once a field hospital or other type of treatment facility is overwhelmed, the overflow of patients could be picked up by the staging team, which would then arrange air transportation for them. Such a staging team, though, would need either to be co-located with the military treatment facility in the theater or be on 24-hour call for rapid deployment in the event of a mass-casualty situation. This would formalize the ability of the Army to deal with such surges and serve as a means of redistributing patients quickly and safely.

Somalia further provided us with a sense of what the low-end requirements are for these types of missions. The U.S. Army is very good at planning the medical support for combat operations; however, it does not have much experience in planning for the lower-intensity end of the spectrum. The patient-level and provider-level analyses of how well demand matched the medical support during this deployment may be useful in better understanding this relationship and in planning future operations.

Another issue associated with planning medical support given variable demand illustrated by Somalia is the need to manage the tradeoff of maintaining relatively rare (and expensive) medical specialties (e.g., neurosurgeons) in-theater versus using them in the peacetime structure. Although the medical need for much of the time was clearly for primary care, the hospital commander faced the problem of keeping his medical staff busy between incidents. This was especially true of the low-density areas of concentration (AOCs) like neurosurgery.

Finally, from a planning perspective, Somalia also illustrated well the tensions that having excess medical capacity in the theater may create in terms of the potential for mission change. In OOTW, having excess medical capacity in-theater is to a certain extent unavoidable, since demand may fluctuate widely and the missions can be highly fluid in nature.[45] If one has excess medical capacity in-theater, demand for services is relatively low, and the medical staff is being underutilized, then there will be a natural tendency to want to use that excess capacity in ways that may go beyond the original mission. In Somalia, this meant providing some care to Somali citizens. This is a key dilemma commanders will continue to face in these operations: how to maximize the efficient use of their medical assets while avoiding taking on an additional mission.

Problems in Treating Civilian Populations

As the case study makes clear, providing medical support to the Somali civilians was not part of the original mission statement, although the medical planning took into account the possibility of the AMEDD being tasked to treat civilians. Still, the medical staff did end up treating Somalis over the course of this deployment (more at some times than others).[46] Partly this was done to fill in the lulls and maintain clinical skills. But it was also because the medical staff and the enlisted personnel saw an overwhelming need in the com-

---

[45]The Army force structure determines what units doctrinally will be required to supply deploying forces. The organization is designed to support a certain level of intensity (e.g., fight a war in Europe), and the force modernization process was aimed at making changes necessary to have units with enhanced capabilities to achieve this. For example, under MF2K, the new combat support hospital (CSH) has much more intensive-care capabilities, and the mix of providers for this hospital will reflect this design factor. In OOTW, however, this means that if a brigade deploys, for example, the level of medical support it may require will not necessarily be at the same level of intensity a brigade may require in a full-scale war. Thus, if a slice of a hospital (e.g., CSH) is deployed for OOTW, then by definition one will have a higher capability than what one might actually believe is the level of intensity associated with this particular operation. The difference between the level of intensity a hospital is designed for and the level of intensity one actually may see in an OOTW is what results in excess capacity in these operations. Interview with COL David Nolan, 4 January 1996.

[46]Although the AMEDD greatly curtailed providing care to the Somalis after the Pakistani incident, such care did continue throughout the remainder of the operation.

munity.[47] Initially, the medical staff was highly motivated to participate in this humanitarian relief effort and provide care to Somali nationals.

Doing so, however, created a host of clinical and operational dilemmas for the AMEDD, with commanders worrying about the potential for mission creep. These dilemmas included:

- What level of medical care to provide Somali civilians in a situation where the host country's standards of care were either rudimentary or nonexistent;

- Whether to treat a medical condition the AMEDD staff knew was treatable by Western standards but not by that of the host country (e.g., diabetes or cancer);

- How to transfer a patient's care to community providers or to a coalition soldier's own country, and what kind of treatment to provide when one knew a patient likely would not have access to follow-up care;

- Whether to provide a level of care the host country might not be able to sustain once the Army departed.

From an operational standpoint, treating Somalis could also tie up beds within the field hospital and deplete the stocks of medical supplies. In the worst-case scenario, casualties could overwhelm a hospital's capabilities if medical supplies were low because of the treatment of civilians. When U.S. casualties increased and beds needed to be made available for incoming wounded, civilian patients might have to be immediately discharged, even if they are not ready for release.

In addition, as Somali support for the mission deteriorated and the risk of U.S. personnel being shot greatly increased, the Army hospital staff began seeing an increasing number of U.S. casualties. These changes led to a great deal of anger and resentment among the medical staff and to considerable debate about continuing to treat

---

[47]Further, nonmedical enlisted personnel brought in Somalis they had encountered who were injured or in need of medical attention.

Somali nationals. A further problem involved the placement of Somali patients on the same wards as U.S. soldiers requiring hospitalization.

The experience of the Swedish hospital in Somalia suggests that if civilians are going to be treated in these types of OOTW, separating civilian and military patients may be a good idea to minimize potential conflicts among the staff and patients. To do so requires having enough ward beds on hand to effect the separation. However, the Army is well positioned to do this, given its modular equipment that is part of Medical Force XXI.

Besides separating civilian and military patients, the Swedes also elected to provide a different level of care to civilian patients—one more in line with the existing medical standards of the host country. Such a policy would be useful for the AMEDD as well.

Benefits of the Somalia Mission for Training

While the Somalia mission clearly presented problems for the AMEDD in structuring its support and in treating civilians, it also had a high training value, especially in the initial phase of a rotation. Specifically, the deployment gave AMEDD personnel valuable training in field medical skills. By providing care to the civilian population, the medical staff were able to treat conditions and operate on wounds they would not normally see stateside, which, in turn, made AMEDD personnel better prepared to deal with the combat casualties they saw later on.

Medical personnel also gained invaluable experience in triage and in handling a mass-casualty situation. The Somalia deployment, for example, proved invaluable in the recent aircraft crash known as the Green Ramp incident. On 23 March 1994, a U.S. Air Force F-16 fighter aircraft collided with a C-130 Starlifter transport aircraft at Pope Air Force Base, North Carolina, as the two aircraft both tried to land on the same runway. At the time, U.S. Army soldiers from the 82nd Airborne Division, XVIIIth Aviation Brigade, and other units from Fort Bragg were standing on the tarmac preparing to board an aircraft for a routine training exercise. The debris and fireball from the collision plowed into the group of Army soldiers, killing 20 and wounding 80 others. The wounded were transported to several military and civilian hospitals, including Womack Medical Center, Fort Bragg. CPT Mango, one of the nurses interviewed, had just returned from Somalia, and made the observation that as a result of the AMEDD's experience during the Ranger firefight incident, the medical staff at Womack Medical Center who had just returned from Somalia were much better prepared to handle this mass-casualty incident. The staff worked well as a team, were able to make the necessary triage and treatment decisions called for, and in general were psychologically better prepared to deal with the kind of injuries they saw from the aircraft crash. On the other hand, the medical staff who had not deployed to Somalia had more difficulty in dealing with this incident.

In addition, this operation provided an opportunity to field test the telemedicine capability—Remote Clinical Consultation System (RCCS). RCCS is an "on-site" system linked to Walter Reed Army Medical Center (WRAMC) that allows providers in the field to consult in real time with medical experts at WRAMC. The telemedicine capability was utilized in the Ranger firefight incident to assist the surgeons in the 46th CSH.

Finally, commanders gained important experience with multinational coalition operations, including the opportunity to evaluate other coalition forces' medical capabilities. All this experience directly contributed to the AMEDD's combat readiness mission.

Chapter Five

CONCLUSIONS

Based on the two case studies presented earlier and additional information on other operations—e.g., the mission in Haiti[1]—we have identified a series of common issues about medical support from which we can draw some generalizations. This chapter discusses these issues in four categories: medical mission expansion and "mission creep"; medical support requirements; problems within coalition operations; and the impact of OOTW support on Army medical readiness and peacetime care.

MEDICAL MISSION EXPANSION

In OOTW, one can expect the medical mission to change as a consequence of the operation itself, thus the need for ongoing mission analysis. In addition, there may be both external and internal pressures for the medical mission to expand—a phenomenon often called "mission creep." In fact, the demand for medical services in OOTW seems open-ended and could consume large amounts of military medical resources, which in turn could undercut readiness for other missions. To manage this demand and to plan for it, the theater commander, subordinate commanders, the JTF Surgeon and his staff, and planners need to recognize forces that exert pressure toward broadening the mission and be able to decide when to resist taking on an additional mission.

---

[1]The original U.S. mission in Haiti was entitled Uphold Democracy; the follow-on UN peacekeeping mission was identified as UNMIH.

Below we identify several ways in which missions are expanded in OOTW: inadequately defined medical missions, mission creep, incomplete mission planning, and changes in the overall operation. We then describe underlying factors that inherently tend to expand such missions.

Inadequately Defined Missions

In some cases, what appears to be mission creep is actually a problem of not knowing the exact mission. An inadequately defined mission results when the scope of the mission, objectives, and specified tasks are unclear and there is wide latitude in how the mission is to be executed. Often, this stems from an absence of guidance from the strategic or interagency levels. For example, during UNPROFOR, the statement of entitlement to care and the population at risk was not well articulated at the outset. The medical objectives that emerged from the interagency planning process were not written at the customary level of detail for a medical mission statement, making it difficult to operationalize.[2] This contributed to mission expansion during UNPROFOR. We suspect this problem applies broadly to all OOTW missions. Without clear guidelines for giving or withholding care, medical providers will tend to react to the immediate need, regardless of the long-term consequences for resources or readiness.

Mission Creep

Under some circumstances the mission may expand not because of vagueness in its definition but because commanders formally depart from their original mission. This is the type of mission expansion most appropriately termed mission creep. For example, the U.S. military may decide to extend the scope of its medical activities to compensate for inadequacies in other coalition troops' medical assets. Or a U.S. commander may decide to allow his medical unit to treat civilian patients and to visit refugee camps to keep the medical

---

[2]The medical mission called for the provision of 30 medical beds and 30 minimal-care beds and care to UN personnel. However, there was no mention of such requirements as an ICU capability, an evacuation policy, etc. Further, there was no legal guidance on whether "UN personnel" included blue card holders, green card holders, contract personnel, State Department employees, NATO employees, etc.

staff busy and to boost morale. UNPROFOR was a good example of the United States electing to provide combat stress support, preventive medicine support, and Echelon IV care to compensate for missing assets and poor quality of care among UNPROFOR troops. However, doing so meant going beyond the scope of the original U.S. medical mission.

Some mission creep is initiated by the Army itself. For example, to date U.S. policy in OOTW has been that if actions of its military inadvertently lead to the injury of a foreign national, then it will treat that person. Thus, when a U.S. soldier brings into a military hospital a foreign national who was accidentally hit by an Army truck or caught in an exchange of gunfire, treating that patient does not expand the original mission. But if an infantry soldier out on patrol comes across an injured or seriously ill Haitian and brings that individual into the U.S. hospital for care, treating that person is an instance of mission creep. These two cases are sometimes difficult for soldiers to distinguish. In both Somalia and Haiti, nonmedical U.S. personnel brought a significant number of injured or sick foreign nationals into the U.S. hospital. At various points during these two operations, the medical staff also undertook elective provision of care to foreign nationals. This occurred partly out of a desire to do what medical personnel are trained to do (i.e., provide care to those in need) and partly to keep busy. However, given how the parameters of the medical mission were written for these two OOTW, providing such care represented mission creep.

Some mission creep is derived from external influences. In recent OOTW, there have been numerous requests from the UN, the State Department, OFDA, other contingents, U.S. and foreign ambassadors, and nongovernmental relief organizations (NGOs) for the U.S. military hospital to undertake the care of refugee or civilian patients. During UNPROFOR, the fact that the U.S. military hospital was not burdened by high patient demand during certain phases of the operation convinced others (e.g., State Department and UN officials) that the hospital was capable of extending its mission to treat refugee children and adults.

When requests for refugee or civilian care come from other countries or from the host nation, the theater commander and CINC have some leeway in their decision. However, the political significance of such requests may not be fully understood at the tactical or operational levels. Further, in the case of requests from the UN, the U.S. State Department, the country team, or other foreign diplomats, the theater commander may seek formal guidance but find that it is not forthcoming.

The presence of coalition soldiers can also prompt mission creep. For example, in UNPROFOR there were external pressures for the U.S. military hospital to assume Echelon IV care and long-term management of certain coalition soldiers. However, some of these pressures were also generated within the military. For example, the medical staff knew that, in some instances, a coalition patient might not receive adequate care if transferred to a local community hospital or repatriated to his own country. The decision to manage the care of such coalition soldiers, beyond the level that would normally be expected, is arguably an example of mission creep that was at least partly under U.S. control.

Dental care is another area with a tendency toward mission expansion, though perhaps not as costly as those above. For example, in recent operations the availability of U.S. military dental care in the theater resulted in a high demand for dental services, particularly among troops from developing nations. Some soldiers sought care for acute conditions (i.e., emergency dentistry problems), whereas others sought care for nonemergent problems (i.e., preexisting dental problems not severe enough to affect their performance). For many such troops, this was the first time they had access to dental care. At the same time, U.S. military dentists and providers may also be imposing their own standards of dental readiness on coalition troops; thus, some of this demand may be provider-induced. Providers often have trouble distinguishing between appropriate versus inappropriate care-seeking behavior. To the extent that the demand for services is provider-induced, then it is within the ability of the U.S. military to control the extent of mission creep that occurs at the delivery end.

Incomplete Mission Planning

Mission analysis can be broken down into specified tasks and implied tasks (derivative tasks necessary to support the specified tasks). For example, during Support Hope, the U.S. military was to assist in the distribution of water to refugee camps in order to halt the spread of cholera. To accomplish this specified task, the implied tasks might include bringing in water purification units or tanker trucks to pump and transport water to the camps, or providing airlift to NGOs so they might bring in their own tanker trucks and pumping units. Implied tasks are where one gets on a "slippery slope" toward mission creep, since often there will be wide latitude in how these tasks are defined and interpreted. Further, implied tasks shade off into other activities generated by an understandable humanitarian impulse to help those in need. Mission planning needs to recognize this tendency and determine implied tasks wherever possible.

In addition, planners need to be able to recognize appropriate planning factors for an operation. For example, in coalition operations, troops from some developing countries tend to have lower levels of medical readiness. If their readiness levels are not considered in the planning process, their needs will probably be recognized later, and the mission will expand to cover them. Some might call this "mission creep," but from another perspective it may simply be a result of incomplete planning.

Changes in the Operation

Sometimes, the nature of an operation is altered because of strategic or political conditions, prompting a change in medical support. For example, in Haiti the multinational force was not on the initial planning horizon; the original plan was to support a U.S.-led invasion force only. It was not until after the initial deployment, as the operation evolved into a "nation assistance" mission, that different patient populations were added incrementally (i.e., island nations' troops, U.S. government employees, Brown and Root personnel, and other civilian contractors).

Thus, during Uphold Democracy, the population at risk evolved from an initial invasion force primarily composed of U.S. troops (with the expectation of heavy combat casualties) into a multinational force engaged in humanitarian assistance and nation-building. With the start of UNMIH, the population at risk expanded once again to include troops from other countries in addition to the United States and the island nations. Such changes in the population at risk brought about an unavoidable result of overall mission changes. Although, strictly speaking, this ought not to be termed mission creep, the increasingly diverse patient population certainly did change mission requirements.[3]

Broad political decisions may also undermine the functioning of a theater's medical system, leading the United States to expand its mission. An example can be found in the loss of Echelon II assets due to the departure of the British medical battalion at the start of the third rotation for UNPROFOR. This led the United States to assume responsibility for some aspects of within-theater patient evacuation. We would argue that filling in a critical gap in Echelon II assets, in this instance, was the result of changing mission requirements.

Underlying Factors Affecting Mission Expansion

Excess capacity in-theater. Underlying many of the issues outlined above is excess medical capacity brought into a theater. By the very nature of OOTW, excess medical capacity is unavoidable given that patient demand often varies widely and planning must consider the worst-case scenario. This excess capacity creates two conditions that contribute to mission creep.

First, because patient demand in these operations tends to be relatively low, medical resources may be underutilized in-theater. Then, in order to keep busy, the theater or medical commander and the medical staff tend to use the excess capacity in ways that depart from the original mission. Thus, having excess capacity in the theater creates a tension at the delivery end and at the tactical level, which encourages additional activities.

Second, excess medical capacity leads authorities outside the normal chain of command (e.g., State Department officials, ambassadors, and UN officials) to pressure the U.S. military hospital to expand their activities for humanitarian purposes (e.g., the treatment of refugees) or in other ways that go beyond the original mission of supporting the deploying force.

---

[3]To some extent it might also be due to an inadequate definition of the population at risk for the multinational force and UNMIH.

Multinational forces. In OOTW involving a multinational force, there are several fundamental design flaws in the structure of the theater medical system. These flaws contribute directly to mission creep. First, no single individual has complete command and control over the entire system. For instance, the UN Force Chief Medical Officer functions primarily in a coordination capacity. Although he oversees the entire theater medical system, he has no authority over any of those elements. One outcome is a disjointed theater medical system that may have holes in the support system, due to a failure by some coalition partners to bring in adequate assets or due to the unexpected withdrawal of key assets. This in turn has led the U.S. military to plug critical gaps in the overall system.

Second, a theater medical system composed of assets from different countries will have wide variance in quality across its components. The U.S. military has responded to this problem by imposing its own standards on that system. For example, in Haiti, the United States was tasked to provide Echelon III care for the multinational force and for UNMIH. At the time the UN mission was being planned, the United States insisted that dedicated aircraft for MEDEVAC be provided. The UN refused on the grounds that it was too expensive and told the United States that if it wanted such craft in-theater for its troops, then the U.S. military would have to bring in its own assets and pay for them (which in fact is what happened). Similarly, given a UN medical logistics supply system that does not meet FDA standards and has poor quality control, the U.S. military compensated by using its own supply system. Thus, insisting on adherence to certain standards in coalition operations has an inherent tendency to expand the mission.

MEDICAL SUPPORT REQUIREMENTS OF OOTW

One can argue that a number of the medical activities undertaken during OOTW are actually similar to those undertaken during wartime. For example, in a Korean MRC scenario, the U.S. Army may be called upon to deal with a large displaced population. During Operation Desert Storm, the U.S. Army was tasked to provide medical support to a large number of POWs. Treatment of complex mine injuries in OOTW also serves to prepare Army, Air Force, and Navy medical personnel for treating these types of casualties during wartime. Therefore, from a medical point of view, it is not so much that supporting OOTW represents a completely different type of operation from supporting combat, but that medical support requirements for OOTW tend to differ in terms of the nature of the patient population to be served, the demand for medical services, and the mix of personnel and units required.

Nature of Patient Population Served

In OOTW, the patient population tends to be broader than those supporting the deploying force and often may be more loosely defined than in combat operations. For example, in addition to U.S. troops, patient subgroups may include troops from other countries, UN or NATO civilian employees, contract personnel, U.S. government employees, and foreign nationals. In addition, because some militaries have a greater proportion of female soldiers and many UN and NATO employees are women, women will form a much larger subgroup than in typical combat operations.

Of course, the AMEDD has been responsible in the past for treating civilian patient populations during typical combat operations. In OOTW, however, treating civilians is more central to the mission, because OOTW often take place in regions where the host nation's health care infrastructure has been destroyed, where there are large refugee populations present, and where there is a humanitarian component to the operation. Thus, treating children and adult foreign nationals may be unavoidable in OOTW.

The presence of a multinational force also broadens the types of patients and treatment demands. Because a number of foreign militaries rely heavily on reservists in OOTW, these troops will tend to be older. There also may be wide variation in troops' predeployment screening, preventive medicine support, and levels of medical and dental readiness. In general, these troops tend to bring more disease, including infectious diseases, into the theater.

Demand for Medical Services

All the above factors suggest that the population at risk in OOTW will tend to show greater variance than in typical combat operations. In other words, instead of primarily supporting relatively healthy, young adults with few chronic medical conditions, medical personnel will support a population that varies in terms of its health status, age structure, gender, types of infectious diseases brought into a theater, and the range of acute and chronic medical conditions that require treatment. As a result, OOTW frequently entail several added classes of demands and can encompass a broader range of activities than do combat operations.

In addition, a multinational force will tend to "up" the support requirements in OOTW, regardless of whether the United States's medical mission is to support U.S. troops primarily or to support an entire multinational force, as was the case during UNPROFOR and in Haiti. For example, linguistic requirements are problematic for a medical unit if tasked to provide medical support to soldiers from a large number of different countries. To indicate the extent of the problem, during UNPROFOR alone the U.S. military hospital was responsible for supporting troops from up to 31 different countries. Further, caring for certain coalition soldiers can lead to resourcing problems. For example, some patients may draw heavily on supplies and require intensive nursing and physician care due to inadequate care received in the field from their own units, poor preventive medicine support, or delays in transport to the U.S. military hospital. To illustrate, during UNPROFOR one soldier dying of AIDS required massive amounts of antibiotics and other hospital resources.[4]

At the same time, although the range of services required may be broader and the support requirements more intensive, patient de-

---

[4]The Argentinean AIDS patient, in particular, is a good example of the kind of resourcing and treatment dilemmas that can arise. This soldier's country was reluctant to have him repatriated and was thus deliberately slow in arranging his transport back home. At the same time, the soldier was drawing heavily on hospital resources. The medical staff knew that if they transferred him to the local community hospital, the local standard of treatment for such patients was to set them aside and let them die. Ultimately, the Argentinean patient was transferred to a community hospital where he eventually died, a decision that caused much anguish among the U.S. medical staff as to what was the correct choice. Local civilian hospitals may also have different standards of medical care, depending on a patient's nationality. In one case, the United States transferred a patient to the local hospital because it could not do anything further for him. This patient was eventually returned to the U.S. hospital, and it was clear upon his return that he had not received proper treatment. This caused the U.S. medical staff to become very reluctant to send any more coalition patients to the community hospitals.

mand in OOTW tends to be relatively low and may fluctuate over the course of a deployment (depending on whether the environment is like Somalia, where the potential for combat is high, or like the Balkans, where the main medical threat is in terms of DNBIs and land mines). As illustrated by UNPROFOR, the season of the year can also affect the level and nature of patient demand in these operations.

Type and Mix of Personnel and Units Required

This variance in demand, in turn, has implications for the types of services and the mix of medical personnel and units required for OOTW. For example, in supporting a multinational force, a broad range of clinical services may be called for, including outpatient clinical services, primary care and internal medicine, ob/gyn care, dental services, physical therapy and rehabilitative services, emergency and trauma care, and surgical and intensive care unit capabilities. As a result, in OOTW the demand for clinical services often represents more of what a community hospital would expect to see than a military hospital (e.g., MASH unit), which is normally geared toward trauma and emergency care.

Because patient demand is relatively low, the size of the hospital required for OOTW also tends to be small (e.g., averaged 60 beds and 120–180 medical personnel for Somalia, UNPROFOR, and Haiti). Rarely do OOTW require an entire military hospital unit; rather, the sections of a hospital need to be tailored (in terms of the number and types of wards deployed and the mix of physicians, nurses, and ancillary personnel) to provide a full range of services. Further, problems in repatriating coalition soldiers may lead to a requirement for a holding ward capability or for minimal-care units. In addition, the demand for preventive medicine, public health services, combat stress support, and veterinarian support can be relatively high in OOTW, particularly when a multinational force is present. Thus, it is not just the military hospital itself but these other elements as well that need to be incorporated to meet the full range of demands associated with these operations.

MEDICAL STRUCTURE ISSUES UNIQUE TO UN AND COALITION OPERATIONS

As we saw in the case studies, the UN's current approach of pulling together medical assets from a number of different countries has led to a disjointed system, in which the different echelons of care may not be connected and in which the quality of assets and medical care may vary widely across the different elements. As a result, in these operations the U.S. military's concept of echelons of care falls apart. Such problems, however, are not unique to the operation discussed in the case studies. Such problems arose during the missions to the Balkans, MFO Sinai, Haiti, and Somalia. Whenever an operation involves a multinational force, a big challenge will be finding ways to link the various levels and different national capabilities to construct a viable theater medical system.

Military medical organizations are expensive to maintain. Relatively few countries can afford to have the full range of capabilities or the same high standards as that of the U.S. military. In addition, a number of countries are currently in the process of downsizing their armed forces, and as they do so, many are shifting the bulk of their medical assets into the reserves. At the same time, some nations that rely heavily on reservists for OOTW (e.g., Canada, Britain) have reported recruitment and retention problems because of the increased frequency of deployments. As a result, when it comes time to assemble the medical support for a multinational force, few nations are willing to commit to providing a hospital (Echelon III care), since doing so not only represents an expensive undertaking, but may also require the mobilization of their reserve forces. In addition, many countries are reluctant to contribute medical assets for OOTW, because it may affect their peacetime military and civilian health care delivery capabilities.[5]

As a result, in coalition operations the U.S. military is often tasked to provide Echelon III care and frequently serves as the backbone of the medical support for a multinational force. However, a key lesson from UNPROFOR was that one cannot isolate an Echelon III hospital.

---

[5]In other words, supporting an OOTW may require pulling physicians, nurses, and other critical staff out of the slots they fill in the peacetime structure.

Problems in repatriating coalition patients can result in mission changes as the U.S. hospital finds itself managing the care of some soldiers far longer than it would normally expect to or providing Echelon IV care. Although there may be a written evacuation policy, in practice it will become situational depending on various coalition partners' capabilities and willingness to repatriate their own soldiers. Such difficulties in repatriation may also tie up beds and hospital personnel, as well as set up the potential for the hospital to be over-loaded in a mass-casualty situation. Further, the U.S. military will not always be able to contract with local hospitals to take on the overflow of coalition patients in a mass-casualty situation or to take on soldiers in need of minimal care only. For example, in Somalia, the local medical infrastructure had been completely destroyed and there were long distances between the Echelon III hospital in Mo-gadishu and fixed facilities in neighboring countries.

Further, in operations involving a multinational force,[6] there may not be dedicated aircraft available for Echelon II. MEDEVAC and medical logistics are especially problematic, because few countries have dedicated aircraft for MEDEVAC. In addition, some forces may have difficulty in keeping their medical activity going, and some may unexpectedly withdraw their medical assets. Some may promise cer-tain assets but in the final analysis not deliver. Finally, some medical units may come into the theater inadequately equipped, supplied, or trained.

Beyond problems with simply holding the echelons of care together, there is often a mismatch between the U.S. military's expectations and those of the UN and its other coalition partners in terms of how the echelons of care get defined, what standards of care will be ad-hered to, what level of medical readiness troops will have, what the scope of the medical mission will be and the population at risk to be served, and what medical policies in terms of host nation support and the treatment of civilians are to be followed. For example, in UN operations the medical logistics system can be unreliable or sub-standard. Finally, as we have noted above, the lack of central au-thority in the UN system leads to coordination problems, and the UN

---

[6]Whether UN-led or NATO-led, a formal combined operation, or an ad hoc coalition operation.

imposes a cumbersome bureaucracy, complex reporting requirements, and additional layers of command and control.

IMPACT ON READINESS AND PEACETIME CARE

In 1995, the AMEDD had personnel deployed to 76 different countries, not including those who were forward-stationed. Although
OOTW tend to require only part of a hospital, this does not mean that
their impact on the system as a whole is minor. Partial deployments
of hospitals affect unit readiness, and certain medical assets have
had a particularly high OPTEMPO in recent years. At the same time,
such deployments affect the AMEDD's ability to deliver peacetime
health care.

Impact on Personnel and Unit Readiness

In general, OOTW tend to affect personnel readiness positively (at
least in the initial phases of a deployment) and unit and equipment
readiness negatively. In terms of personnel readiness, the training
value of recent operations has been relatively high, though this value
is concentrated within the first few months of a deployment. In fact,
one could argue that the greatest training value of OOTW occurs
during the first rotation and lessens during subsequent rotations,
since later hospitals tend to fall in on the first rotation's equipment
and just rotate their personnel into the theater. An exception is
treating mass-casualty incidents or complex trauma cases (e.g., land
mine injuries) and conducting triage, activities that can take place
during any rotation. In addition, the training value of OOTW tends
to be greatest for commanders and their core staff and for certain
medical specialties (e.g., dentistry, preventive medicine, and orthopedic surgery). However, this value will depend on the individual
operation and (as seen in the two case studies) may be offset by the
underutilization of the medical staff during a deployment.

OOTW tend to require only sections of a hospital tailored to meet the
demands of a specific mission (e.g., in support of the UN multinational force (MNF) in Haiti, only half of the 47th Field Hospital's assigned personnel were deployed); as a result, such partial deploy-

ments can degrade a TOE medical unit's readiness posture.[7] Although one may assume that the remaining portions of a medical unit are available to support another deployment, this may not be true, since if critical elements are taken (e.g., command and control element, or the only x-ray section of a TOE hospital), then the remaining sections will be unable to undertake an additional mission.

Such partial deployments are potentially devastating, primarily because of the equipment densities of these units. Table 5.1 shows the medical equipment densities for various types of Army hospitals in the current inventory. In the case of a field hospital, for example, if a third of the unit is deployed, the equipment requirement to support it could entail sending the hospital's only complete x-ray, central material service (CMS), pharmacy, operating room, blood bank, laboratory, medical maintenance, or occupational therapy/physical therapy (OT/PT) sections. However, for a general hospital, sending one x-ray section still leaves the hospital with a second section that could be used for a second deployment.

Partial deployment of hospitals also may affect training, since the remaining part of the unit cannot train if critical equipment is deployed.

The deployment of low-density specialties would probably have little impact on the readiness posture of a TOE hospital, because these medical specialties are PROFISed to a hospital and not permanently assigned to it; thus, another individual within the system could be tapped for the second deployment. Instead, the impact would be more of a systemwide problem if there are insufficient numbers of these individuals in the areas of concentration most needed for OOTW.

However, the AMEDD personnel who may significantly affect a unit's readiness are those permanently assigned to the command and control element of a hospital (i.e., the executive officer, operations officer, company commander, and communications sergeant—the individuals crucial to operating the unit). If this element is being utilized

---

[7]Although we did not examine the impact of specific deployments on unit readiness, the following summary is intended to serve as a basis for considering what the potential effects may have been, given how the AMEDD has supported recent deployments.

Table 5.1

Medical Equipment Densities for Army TOE Hospitals

Hospital Type	X-ray	CMS	Dental	Phar-macy	Oper-ating Room	Blood Bank	Lab	Med. Maint.	OT/PT
CSH	2	4	3	1	4	1	1	1	1
General	2	4	3	1	4	1	1	1	1
Field	1	1	3	1	1	1	1	1	1
MASH	1	1	0	1	1	1	1	1	0

for a partial deployment of the hospital, then the remaining hospital's readiness posture will have been reduced and training compromised.

An additional factor affecting unit readiness is the amount of time it takes to bring a TOE hospital back to its full capabilities following a deployment.[8] Depending on how long and intensive the medical requirements were for a given mission, the recovery period may take up to several months.[9] For example, in the case of the 212th MASH unit (the hospital that undertook the first rotation in UNPROFOR), the recovery process was more extensive because the hospital was required to leave its equipment in place for subsequent rotations to fall in on. In this case, the unit had to receive completely new or refurbished equipment, a process that can easily take two to three months.[10]

---

[8]This issue often only applies to the first unit that deploys if subsequent rotations fall in on the initial medical unit's equipment.

[9]In a normal overseas deployment, the recovery process typically involves seven steps: (1) pre-redeploy equipment and supply inventories; (2) tear down and pack up; (3) transport equipment and supplies to the port; (4) redeploy medical personnel; (5) receive equipment at the home station; (6) perform maintenance inspections of equipment; and (7) order replacement medical supplies.

[10]It would entail identifying equipment shortages, ordering, scheduling the DEPMEDS field team, receiving new equipment, inventorying equipment, reordering supply shortages, performing maintenance checks on new equipment, and training on the new equipment.

In addition, some units have particularly high OPTEMPO (e.g., area support and medical logistics battalions) associated with OOTW. Contributing to this is the fact that beneficiaries must be served in both CONUS and Europe. To illustrate, because two of the four existing medical logistics battalions in the active-duty structure are in Korea and Europe, FORSCOM has only two medical logistics battalions it can access for OOTW.[11] The hospitals have a similar (although lesser) problem. Currently, Europe has three fixed facilities, all of which are TOE units. If one of these TOE hospitals were deployed, it would mean that a substantial number of the hospital's medical staff would have to be backfilled with PROFIS fillers from CONUS to maintain beneficiary care in that installation's facility. In turn, however, the CONUS facilities from which the PROFIS fillers were pulled would themselves need to be backfilled.

Impact of OOTW on Peacetime Health Care Delivery

In previous unpublished RAND research we reported analyses of the impact of recent OOTW on peacetime health care. Here we summarize the important observations and present some recent data from the Haiti deployment. Since the financing of these operations can affect the AMEDD's OPTEMPO and its ability to maintain peacetime health care, we also discuss efforts undertaken by the AMEDD to minimize the financing impact of recent operations.

Although the costs of undertaking OOTW are relatively small compared to the total Army budget, their impact on peacetime health care can be substantial if concentrated in a few programs or installations. Further, these operations tend to whittle away at all levels, including loss of personnel, unreimbursed costs incurred in support of an operation, unfunded programs, or lost training opportunities.

The impact of OOTW on beneficiary care has been a function of the size of the military treatment facility (MTF) and the amount of draw

---

[11]These units are the 18th MEDLOG Battalion in Korea, the 147th MEDLOG Battalion (Rear) at Fort Sam Houston, the 32nd MEDLOG Battalion (Forward) at Fort Bragg, and the 226th MEDLOG Battalion (Rear) at Pirmasens, Germany.

on a facility's medical personnel. The loss of PROFIS personnel[12] for some deployments, for example, has been felt primarily by the smaller installations such as Fort Drum's small medical activity (MEDDAC), which supported the 10th Mountain Division's deployment to Somalia. However, the loss of PROFIS personnel can also be felt by some of the large medical centers (MEDCENs) because of problems the AMEDD has had in relying on reserve volunteers to backfill PROFIS losses sustained by a medical center. To illustrate, for Uphold Democracy the backfill requirement for this deployment was for 81 Individual Mobilization Augmentees, of which 48 were to be physicians.[13] However, only 24 such augmentees were identified (just five of whom were physicians), indicating a significant shortfall in meeting the requirement through volunteerism, particularly in terms of doctors. As a result, Womack Army Medical Center at Fort Bragg, which sustained the largest PROFIS losses because of the deployment of the 28th CSH to Haiti, had its MEDCEN's capabilities degraded (at least temporarily). Specifically, initially three operating rooms had to be closed and the number of elective surgeries reduced, and the North Atlantic Health Services Support Area (HSSA) had to cross-level critical personnel (i.e., bring in active-duty personnel from other MTFs) to help Womack maintain beneficiary care.

The financing impact of these operations on OPTEMPO will also be felt in terms of lost training opportunities and unfunded programs. Further, the impact will be related to the timing of an operation during the course of a fiscal year. Because contingency operations are not programmed for, funds must be diverted from other purposes. So if an operation occurs early in the year, funds can be borrowed from subsequent quarters with the hope of recouping those expenses prior to the end of that fiscal year. However, operations that occur late in a fiscal year have little chance of recouping their expenses in time to utilize those dollars for their intended purpose.

---

[12]PROFIS is the Army's mechanism for transferring individual professional personnel to deployed or forward-stationed units when needed. For example, during an overseas operation individual physicians may be transferred from a CONUS hospital to a deployed unit. Often the vacant slot in CONUS is then "backfilled" by another physician, perhaps one from the Reserve Components.

[13]That is, the number of reserve medical personnel needed to backfill PROFIS losses for this deployment was 81.

During fiscal years 1993 and 1994, the AMEDD was successful for the most part in minimizing the impact of these operations on beneficiary care. It did so by several means. First, the AMEDD spread the PROFIS requirement across the system in supporting some operations. For example, for the third rotation of Operation Continue Hope (Somalia), the USAMEDCOM pulled a few PROFIS physicians and nurses from a number of different MTFs across CONUS to staff the 46th Combat Support Hospital and to backfill losses, rather than drawing all the hospital's PROFIS requirement from a single MTF.

Second, the USAMEDCOM also tried to minimize the financial impact of these operations by reimbursing MTFs directly to ensure that no program would go unfunded within a fiscal year. That is, the MEDCOM itself, rather than the individual MTFs, took the risk of coming up short at the end of the fiscal year. The 7th MEDCOM in Europe was able to minimize the impact of UNPROFOR on peacetime care by charging all unreimbursed costs to USAREUR's readiness (P2) account.

With the AMEDD downsizing and the number of MEDDACs and MEDCENs decreasing, the potential of these operations to impact peacetime care will likely increase (particularly in Europe), since the number of troops may decline but the beneficiary population does not necessarily drop at the same rate. The Office of the Assistant Secretary of Defense (Health Affairs) has recently become concerned about the impact of these operations, particularly the effect of recent overseas deployments on OCONUS beneficiary care. As a result, all three services were asked to submit a plan on how to minimize the impact of future OOTW on beneficiary care.

We also found that the AMEDD has tended to absorb a number of the direct and indirect costs associated with these operations. As we have discussed, the AMEDD often serves as the backbone of the medical support in terms of providing medical expertise, equipment, personnel, or supplies, often with little hope of recouping many of these expenses. Further, in UN-led operations, U.S. military medical units have tended to rely on their own service's supply system rather than the UN's (again without reimbursement). As seen in Haiti and UNPROFOR, differences in standards between the UN and the U.S. military medical organizations can also lead to the unofficial assumption by the United States of different echelons of care (e.g.,

MEDEVAC). Because U.S. standards tend to be higher than the UN's and because the United States always ensures a stand-alone capability to take care of its own troops, this has meant that the United States essentially augments the UN's medical assets and the theater medical system itself in these operations; in doing so, it also subsidizes the medical component, since these efforts come out of U.S. funds, not UN funds.

This problem is exacerbated by the fact that reimbursement from the UN tends to be slow and may not fully cover expenses incurred in support of an operation. Reimbursement from other coalition partners also can be uncertain.

Chapter Six

FUTURE DIRECTIONS

This chapter describes our recommendations for future directions that the Army should take to improve medical support for OOTW. In some cases we suggest specific changes, for example in defining mission scope and planning for particular kinds of patient populations. In other cases we can only raise issues that seem to be centrally important but that must be resolved by policymakers in the Army and other government and international organizations.

Because many of the issues and problems identified in this study need to be addressed at the strategic, operational, and tactical levels, we discuss three classes of possible Army actions to deal with OOTW: (1) actions the Army and the AMEDD can undertake when their role is clearly defined; (2) actions they can undertake in the absence of clear guidance from higher authority; and (3) actions they may undertake to influence the strategic planning process. We also make recommendations for training and equipping forces to meet OOTW demands.

DEFINING AND SCOPING THE MEDICAL MISSION

Factors Driving Mission Expansion

As we described in detail in Chapter Five, there are strong pressures toward expanding the medical mission as an operation unfolds. To some extent the Army's activist "can-do" approach toward any mission amplifies this tendency. Unfortunately, neither the Army nor the other military services have the medical structure and resources to tackle all of the medical needs in most areas of the world. Without more clearly defined limits on the medical mission, the demands will quickly outstrip U.S. capabilities and may backfire if other countries' and relief organizations' expectations of medical support are not met.

Up to now, the United States has not articulated a national medical strategy that defines the objectives and medical ROEs for OOTW. To a considerable degree, this is due to the continuing belief that the medical mission is limited to its combat service support role. However, in OOTW medical tends to play a more central role. In operations involving disaster relief, humanitarian assistance, or refugee populations, the medical mission may actually be broader than the basic workload of supporting the deploying force. The U.S. strategy for OOTW needs to recognize that many factors push the services toward accepting a larger mission in these operations. These factors include:

- **Needs of civilian populations.** The local population often has evident medical needs. In addition, refugees may be present, and existing medical infrastructure may be destroyed or inadequate.

- **Needs of coalition partners.** A broader set of treatment demands arises among soldiers from other nations and civilian employees of the UN, NATO, or contractors. Some coalition troops may utilize the theater medical system in ways it was not intended. Coalition partners' own medical assets may be inadequate for the mission. And if a key medical asset is not available (e.g., air evacuation), the United States may feel obligated to provide it.

- **Demand induced by U.S. actions.** The U.S. informal policy described as "if we hurt them, we fix them" leads to involvement with civilian populations in any event. In addition, U.S. soldiers may bring in sick or injured civilians on their own.

- **Ethical and professional considerations.** Medical personnel have a professional orientation that implies an obligation to help with urgent medical problems and an understandable desire to respond to medical need, regardless of the official mission.

- **Excess capacity.** In OOTW, excess medical capacity is unavoidable to a large extent given wide fluctuations in demand and the fluid nature of these operations. As a result, inevitably, the in-theater medical facility will be underutilized at times. This available but unused supply tends to stimulate demand. Providers also want to continue practicing their specialties to maintain their clinical skills.

- **Outside requests and influences.** The UN, coalition partners, foreign ambassadors, other U.S. agencies, and the State Department often urge the Army to expand medical services and to utilize any excess capacity for purposes other than the original mission. Coalition partners also may define a broader medical mission and set of objectives for themselves, creating a disparity between theirs and those of the United States which in turn pushes the United States in a similar direction.

Defining the Medical Mission Clearly

Ideally, what is needed is a clear definition of the scope of the medical mission, particularly about how to treat civilians and coalition forces. Based on our case studies, in this section we outline five important elements that the Army, the DoD, and the U.S. government should consider in scoping the mission and discussing options:[1]

- U.S. objectives in providing health services;

- The desired end state for medical support in a region;

- Delimiting populations eligible for services;

- Civilian patient care, transfer and evacuation;

- Relations with other health care providers in the operation.

Objectives. In undertaking an OOTW, the United States should determine whether the objective is to raise the general level of health in a region among the civilian population, or to respond only to the emergency medical needs of a particular crisis. If the policy objective is to improve the general health level, then the Army might deploy, for example, small medical teams to work with local providers to

---

[1]Another important element is U.S. objectives in supporting a multinational force as part of a coalition operation. Because these operations have a number of features critical to scoping the mission, we discuss those issues in more detail and outline a set of recommendations in a later section.

provide preventive medicine support,[2] conduct public health education, or reestablish local clinics and medical facilities.[3] Another objective might be to generate goodwill—which can be a valuable element in U.S. foreign policy. However, this may involve still greater resource commitments. If the United States chooses to provide medical support for this purpose, then it should say so up front, resource it, and buy it inexpensively. We would further argue that only in the case of responding to a specific crisis would it be appropriate to utilize military medical assets to provide civilian care, and then only when the military brings unique capabilities to the mission.

End state. The United States should establish what level of care will be provided to civilians (for example, patients injured by U.S. actions or brought in by U.S. soldiers). We would argue that the United States should not be in the position of providing a level of health care to civilians in OOTW that cannot be sustained once the military departs and that may create unrealistic expectations from the host nation and the NGOs. Further, from a health policy standpoint alone, it does not make sense for military hospitals to treat civilians or provide them with state-of-the-art medicine when often the far greater medical need is for public health and preventive medicine services. In addition, a U.S. military hospital is limited in the range of services it can provide in the field and can sustain over the long run. Instead, U.S. policy in providing medical care to foreign nationals ought to be to treat civilians in-country, at a level that the local health care system can support, and to not go beyond that.

Populations eligible for services. The United States needs to articulate an operational definition of entitlement to care that different patient subgroups are to receive. In these operations, in addition to U.S. troops, other eligible groups may include coalition partners and those connected with the force, such as civilian contractors and civilian employees of the UN or NATO. Presumably, not all groups of patients will (or should) have the same access to medical care as the force itself.

---

[2]For example, stopping the spread of cholera through the refugee camps in Rwanda required immediate action in terms of the distribution of clean water to the camps.

[3]The Special Operations Forces provide a good example of the successes these types of medical missions have had in the past.

For example, the U.S. military may be willing to treat civilian contractors it specifically employs, but not necessarily UN or NATO civilian employees. It may contract out their care, instead. Or it may provide emergency services only to these patient subgroups and require that they be transported out of theater for more definitive care. The military needs to insist on obtaining a clear legal definition of the population to be served.

Patient transfer and evacuation. The medical mission statement should determine the policy and procedures for transferring civilian patients from U.S. military hospitals to local facilities. The medical policy during the Haiti operation was to transfer civilian patients' care to local hospitals as soon as possible. In the Balkans, the U.S. military was fortunate to be in an environment where local community hospitals were available to receive transferred civilian patients. However, even then, if the U.S. hospital had taken casualties in sufficient numbers that required the release of such patients (so beds could be made available) while local community hospitals were also filling up with civilian casualties, the U.S. military might have found itself in the position of having to discharge patients who were not ready for release. In such cases the resulting adverse publicity alone could undo all the goodwill generated in the first place by treating refugees, and this would play at the highest policy levels.

Relationship with other providers. It may be possible to use civilian contractors to provide the medical support for a multinational force in operations where there are no U.S. troops on the ground. This is an expensive option, however, and it was resisted by the UN for Haiti on reasons of cost alone. Recently, the United States considered contracting out the medical support for the UN peace operation in Haiti (UNMIH), but decided against it as being too expensive, since securing civilian physicians and nurses to come into high-risk areas and treat high-risk patient populations would require hefty compensation.[4] Further, contractors may still require U.S. military support in terms of security, logistics, and airlift. And historically, when the U.S. military has utilized civilian contractors for medical support, it has had difficulty controlling them, ensuring quality of care, and get-

---

[4]Interview with COL Snyder, executive officer, Office of the Army Surgeon General; Health Care Operations Conference, San Antonio, TX, June 1995.

ting them to sustain medical support in the way that regular Army units can.

The U.S. military could also contract with NGOs to take over the care of certain groups (such as coalition soldiers or displaced persons).[5] For example, the United States could triage to an NGO those coalition patients who are difficult to repatriate or who can be moved in the event of a mass-casualty situation in order to free up beds. However, the use of NGOs will not work in all situations. Realistically, few will have these kinds of capabilities, many will be unable to sustain their response over the long run, and they may be in-theater for unpredictable lengths of time. The nationality of a coalition soldier also may influence the willingness of an NGO to take on his care. Some relief organizations may wish for the U.S. military to completely assume certain functions, whereas other organizations may only require medical logistics support, transportation, security, or medical teams to assist in the implementation of a program. Further, many relief organizations may be unwilling to have a formal association with the U.S. military out of concerns that they might be viewed as tools of American foreign policy.[6]

For OOTW, the Army should seek to get a national policy that articulates medical objectives and medical rules of engagement covering the areas discussed above. Although formally such mission definition may be the province of other government authorities (e.g., the Joint Staff or the State Department), the Army and the AMEDD need to become more proactively involved in the strategic planning process. Up until now, issues regarding the medical support and the medical mission itself have had little visibility at the Joint Staff or strategic level. Yet by the very nature of these operations, medical often plays a more central role in OOTW. In addition, the Army and

---

[5]Successful examples of collaboration with NGOs can be found. For example, during Provide Comfort the United States worked well with UNICEF in helping it implement an immunization program for Kurdish refugees. Key to the success of this undertaking was the fact that UNICEF had a limited, clear set of objectives for its operation.

[6]On the other hand, as was the case in Somalia and the Balkans, some NGOs may be concerned that U.S. Army medical units may take over their mission (i.e., compete with them) and so do not want the U.S. military to provide care to refugees at all. Others may be concerned that Army medical units may raise the level of expectations in the theater to one that the NGOs or the host nation cannot sustain upon the departure of the U.S. military.

the AMEDD will often provide the bulk of medical support.[7] Thus, the Army has a big stake in ensuring that missions are defined in executable ways. The Army can exercise influence, for example, through its representatives on the Joint Staff and through the role of the Army Chief of Staff in the strategic planning process.[8] The Army Chief of Staff, through his position as a member of the JCS and as an adviser to the National Command Authority, will have the authority and the ability to raise the Army's concerns about the medical mission and obtain clarification on objectives and medical rules of engagement at the strategic level. If the Army does not play more actively in this process, it will be limited to providing input only with respect to its Title 10 responsibilities and may continue to be faced with unclear or unsustainable medical objectives in future OOTW.

Launching the Medical Mission as Defined

Once a mission definition has been agreed upon for an OOTW, it is essential to start off the operation on a strong footing. Our analysis of the case studies suggests two key areas in which the Army and other military forces should take early steps to establish mission limits.

Broadcasting the medical mission. The theater and subordinate commanders need to do a better job of broadcasting the U.S. medical mission to the host nation, local health officials, other troops, relief agencies working in a region, and the press—as one means of averting misunderstandings as to the U.S. mission and medical policies for a given operation. This should include the definition of the patient population to be served and the types of services to be provided to different subgroups. In this way, the United States may clearly articulate medical mission parameters (e.g., treatment of civilians or a multinational force) and avoid unrealistic expectations.

---

[7]The Army normally has executive responsibility for combat service support in a theater of operations. In addition, the bulk of the U.S. military's medical assets reside in the Army.

[8]The AMEDD recently has had a two-star general officer assigned to the J-4, which ought to serve to increase the visibility of medical issues at the Joint Staff level.

Negotiating the workload with other parties. The burden of working with the local government, local community, and various relief organizations will fall on the services. This suggests that military medical units may need to become more involved in the interagency planning process at the tactical level. Although a civil-military operation center (CMOC) can have a medical cell established within it for this purpose, this rarely occurs. Further, we argue that the officers staffing a CMOC will tend to be too junior to deal with the political aspects of the medical mission. What is needed instead is a senior military medical officer with the experience, authority, and visibility to effectively negotiate the coordination of care of civilians, the transitioning of their care to community hospitals, and the range of activities the U.S. military medical units will undertake in assisting the local community.

To facilitate such negotiations, we recommend that the services or DoD attempt to form a closer working relationship with such organizations as the United Nations High Commission for Refugees, the International Organization of the Red Cross, the World Health Organization, the International Organization of Migration, and other key relief agencies to coordinate the provision and transitioning of care of civilian patients. The JTF Surgeon and his staff are the best candidates to assume this role, as was done in Haiti. This officer would also be responsible for advising the theater commander about the type of assistance required by the host nation and relief community, and for interpreting which activities fall within the scope of the mission.[9]

There are several recent examples that can serve as a template for future operations. During UNPROFOR, the Air Force worked closely with the above organizations to coordinate and establish guidelines for the selection and treatment of refugee adults and children by the U.S. military hospital in Zagreb. This included assigning responsibility for patient evacuation and delimiting the operating parameters by which the military hospital would provide refugee care.

---

[9]This has implications for the organization of the medical support to allow the JTF Surgeon or medical unit commander and his staff to undertake such activities. For example, in Haiti, the JTF Surgeon had a small headquarters staff assigned to him to accomplish this.

During the operation in Haiti, the JTF Surgeon served as the key medical interface with the Haitian government (e.g., the Ministry of Health), various relief agencies, the Pan American Health Organization, and such U.S. governmental organizations as USAID. In this way, the U.S. military was able to minimize the civilian care undertaken by its hospital and establish a mechanism by which injured or ill Haitians could be readily triaged and transferred to local hospitals.

BUILDING A SOLID BASE OF OOTW EXPERIENCE AND KNOWLEDGE

The Army and the other services now have a substantial amount of information about how to execute medical OOTW missions, given their recent experiences in Somalia, Rwanda, Haiti, and the Balkans. However, this information is not widely disseminated even within the medical commands, and the medical support implications are even less well understood in "line" organizations (which normally command the medical elements). Under these circumstances there is a risk that future operations may be planned and launched without benefit of the valuable experience already accrued. Therefore, we suggest several actions the Army and the other services might take to preserve and build the base of knowledge about conducting OOTW medical support.

First and foremost, the Army and the AMEDD need to understand what these operations are about, their complexities, and how they differ from combat operations. This is critical for planning and tailoring medical support. It is also essential for articulating policy on treatment of civilians and coalition soldiers, and for defining and operationalizing the medical mission's scope. Although policy guidance on these issues should ideally come from the DoD or the State Department, in the absence of such guidance the Army and the AMEDD need to have a plan in mind. For example, the Army may need an evacuation plan for civilians in case an Army hospital is tasked to provide care to foreign nationals.

Pooling the Services' Information

Of the three services, the Army most often serves as the backbone of medical support in OOTW. Given this, the AMEDD may want to take the lead in convening a conference of commanders and medical staff from all three services, as well as from key coalition partners, to discuss UN and NATO medical issues and how to better support coalition or combined operations. Some of the important issues the U.S. military and its coalition partners will need to address include national differences in mission definition, medical policies, troops' levels of medical readiness, and quality of medical assets.

Relations with NGOs and International Organizations

It is clear that the Army and other services do not yet understand how to interact and coordinate with civilian relief agencies and UN entities also involved in OOTW health care delivery. The Army needs to establish more effective methods for communicating with these organizations. Among NGOs, the Army and the AMEDD should identify which organizations it can effectively work with (e.g., those with well-defined and limited missions and adequate resources) and establish an ongoing liaison with these organizations that can carry through on a number of operations. The AMEDD could also establish a list of officers to serve as points of contact for these organizations. At the same time, it is equally legitimate for the Army and the AMEDD to identify those civilian organizations it may not wish to get involved with. Such steps would establish continuity that would facilitate use of NGOs in future OOTW. The example from Haiti, where the JTF Surgeon served as the key medical interface with the Haitian government, various U.S. governmental organizations, and the Pan American Health Organization (PAHO), is a useful template for future operations.

Individual Education and Training

Key to making the above structure work well for OOTW is the education and training of AMEDD officers and enlisted personnel. As noted in Chapter Five, medical commanders are instrumental in controlling mission creep and in clarifying mission definition. In the training environment, a number of these issues may be addressed.

Clearly, OOTW have a large political element. However, many AMEDD officers and enlisted personnel who are deployed on these missions have had little experience with dealing with these issues at their level of career development. For example, many AMEDD officers are not used to working with other countries' militaries, which may have a different set of political objectives, mission goals, and medical policies. Army and AMEDD officers also are not used to handling direct requests from ambassadors (U.S. or foreign) or from the UN, or dealing with policy issues at the tactical level.

There is a need, therefore, to educate officers at appropriate levels about political issues, UN issues, and coalition and combined operational issues that may arise during the course of these deployments.[10] At the individual level, information on OOTW needs to be incorporated into medical officer professional development courses. For example, in the Officer Basic Course and Officer Advanced Course, a basic introduction to OOTW should occur, including a review of lessons learned from recent operations and after-action reports as well as participation in problem-solving exercises. The Command and General Staff College could provide a forum for holding discussions on the medical support requirements, public health issues, and other problems medical units face in OOTW. The Army War College curriculum could include coursework on planning and leading these operations and how policy and political issues may be addressed by commanders.

Training Exercises Integrating Medical and Nonmedical Units

The AMEDD needs to become more proactive in educating line officers, in addition to medical officers, about medical issues that may arise during OOTW. During an operation, it will be up to AMEDD officers to advise a line commander on the implications of his decisions in terms of the medical and overall mission. For example, if a theater commander decides to allow his MASH unit to treat civilians, then he needs to be made aware that providing such treatment might tie up beds or medical personnel and use up critical medical or blood supplies. Further, in the event of a mass-casualty situation, civilian patients might have to be released unexpectedly without assurance of an available local hospital or clinic to receive them.

---

[10]See Appendix C for a summary of the current initiatives in OOTW training and education for AMEDD officers and medical units.

One way to accomplish this would be for Army medical units to become more involved in collective training for OOTW at the Joint Readiness Training Center (JRTC).[11] It is in such a training environment that line officers and Army medical officers could hash out medical decisions associated with OOTW prior to a deployment, rather than rely on ad hoc decisionmaking in the theater. Further, it is here that providers and commanders may receive training on interpreting an operation plan, developing a tactical plan, and making the kind of clinical and command decisions they might face in a Somalia or Bosnia or Haiti scenario. It will be up to the AMEDD to articulate a future training strategy that exposes Army medical units and other types of units to the medical support and public health issues associated with OOTW.

Training Involving Medical-Unique Issues

At the individual or tactical level, an important question is how to train an Army physician or nurse to respond appropriately in these operations and help avert the tendency toward assuming an additional mission. As described in Chapter Five, the medical staff themselves may inadvertently contribute to mission creep in several ways. First, a physician may "pull" into the theater medical equipment and supplies he is accustomed to using in a peacetime setting (especially, the specialty-trained providers) in order to provide state-of-the-art medical care. This problem is not unique to OOTW, but is part of the fundamental dilemma the AMEDD faces in training medical personnel to adjust to the differences between operational and peacetime medicine.[12] Second, the medical staff may face difficult decisions in the field, such as ethical treatment dilemmas that may arise in dealing with multinational forces. Physicians and nurses need to understand how their actions may inadvertently expand the mission or why certain decisions at the operational or strategic levels were

---

[11]That is, Army medical units need to train with those units and troops they are going to support.

[12]In fact, one may argue that in peace operations and OOTW, in general, there should not be much difference between peacekeeping and peacetime medical care (as opposed to wartime care).

made to contain the treatment burden. Further, the medical staff needs to be given realistic expectations about what they may be called upon to do in these operations.

In addition to JRTC training, one way to accomplish this would be for the AMEDD to undertake medical-unique training for OOTW at Camp Bullis. Such training could include, for example, dealing with ethical and treatment dilemmas that may arise in supporting a multinational force. These issues need to be addressed in normal peacetime training; the predeployment preparation phase does not allow enough time to handle them.

The AMEDD also may want to articulate humanitarian ROEs for enlisted and nonmedical officers (although enforcing ROEs is a command responsibility). As we have noted above, mission creep in recent OOTW was exacerbated by nonmedical personnel bringing injured civilians into the U.S. military hospital for treatment.

FLEXIBILITY IN PLANNING

As shown in the two case studies presented above, OOTW tend to be fluid, resource intensive, unpredictable along five or six different dimensions, and characterized by rapidly changing mission requirements. These characteristics make OOTW difficult to plan, requiring that the medical units and the mix of medical personnel be tailored (and sometimes retailored) to meet the demands of the specific mission. This places a premium on flexibility in planning, ongoing mission analysis, and adaptation (such as anticipating changing mission requirements and task organizing in response). To operate in this adaptive way, planners need to understand how to match the right personnel and units with the mission.

Identifying Determinants of Medical Demand

From a medical standpoint, the critical distinction in OOTW is not whether it is a peacekeeping, peace enforcement, or humanitarian assistance operation. Rather, the key determinants of the medical support requirements include the following:

- Whether the United States is acting unilaterally or whether a multinational force is involved and the level of support the United States has been tasked to provide that force;

- To what extent refugee or displaced populations are a factor;

- Whether there is a humanitarian component to the operation;

- Differences in medical readiness among coalition troops;

- Degree to which the host nation's medical infrastructure may have been compromised;

- Variation in coalition partners' medical assets.

Recent experience should permit military planners to characterize the mission in terms of such determinants. These factors are important to understand because they drive (a) the population at risk, (b) the nature of patient demand for medical services, and (c) the nature of medical resources needed to deliver services. For example, we know that lack of preventive medicine assets and poor quality of care by some troops' medical teams mean that a U.S. Echelon III hospital may receive more patients who are complicated to treat and more resource intensive in these operations. Although it is not possible to plan for all eventualities, considering the above determinants should help the Army do better at predicting OOTW support requirements.

Advance Assessment Needs

A key element in tailoring the force is conducting an advance assessment. Planners at the operational and tactical levels need to understand the type of expertise required for OOTW and assemble the right kind of assessment team (composed of not only physicians, but also preventive medicine officers and community health nurses, for example). A relatively small number of AMEDD officers have had extensive experience in tailoring the medical support for these types of missions. It will be important to spread this experience across the AMEDD and to make sure experienced people are involved in the planning process.

Planning for More than the Combat Support Role

Line planners need to consider not only the medical support requirement for a deploying force, but also what medical force may be needed to achieve specific medical objectives and ensure the mission's overall success. For example, in OOTW involving disaster relief, humanitarian assistance, or refugee populations, the medical mission will be broader than the basic workload of supporting the deploying force. The real thrust of the AMEDD's workload may be in terms of health support to the host nation.[13] However, because the Army and DoD continue to view the medical mission as limited to traditional combat service support, at times they have used the wrong basis for planning, requiring combat medical units to undertake a wide range of activities they were not intended for. This has led to a mismatch in the medical force provided and at times to the inappropriate and inefficient use of medical assets.

There are other unique features of OOTW that make it necessary to consider additional planning factors. For example, planners need to recognize the civilian patient demand units may face. A political reality of OOTW is that the U.S. military will sometimes be tasked to provide care to civilians, whether it is part of the official medical mission or not. As we have noted above, such care will likely include local civilians (especially in emergency situations). Depending on diplomatic or other pressure, it may also include employees and contractors of the UN or NATO, other foreign nationals, and of course soldiers from coalition forces.[14]

Patient and Provider Databases

One way of improving the accuracy of planning the medical support is to standardize patient and provider databases on deployments across all three services. Such data are needed to better understand the nature and level of demand associated with OOTW, and to assess

---

[13]Line commanders often do not appreciate this. Several line commanders who have returned from OOTW told us that what they found they really needed to know more about were medical and public health issues.

[14]Since many countries rely on reservists for OOTW, from a clinical standpoint their forces will look like civilians.

the effectiveness of new technologies such as telemedicine in a field setting.

BUILDING A ROBUST AND FLEXIBLE STRUCTURE

The current challenge to the Army and the AMEDD is to create a medical structure sufficiently flexible and tailorable so that one can readily adjust to one's position along the spectrum of conflict. Because OOTW tend to encompass a broad range of medical tasks, but require less total capacity than combat missions, it does not make sense for the AMEDD to construct new structure for these operations. Further, one needs to be careful not to create extra structure or put into place such a large structure that the system itself may become more unwieldy and inflexible in the process. The key is to build a robust and flexible structure that can respond to a broad range of demands.

Modular and Tailorable Structures

The modular concept of Medical Force XXI (formerly known as the Medical Reengineering Initiative) has features well suited for providing the kind of flexibility the AMEDD requires. The modular structure should allow the Army to tailor its medical units to meet the varied support requirements of OOTW. As noted in Chapter Five, since OOTW tend to require only parts of a military hospital rather than whole units, it will be important to ensure that a partial deployment of a TOE hospital does not significantly affect the readiness posture of its remaining sections.

Contagious Diseases

An exception in terms of new structure is the need for a deployable isolation ward capability. The isolation of contagious patients is difficult to achieve in a tent environment, and certain diseases, such as tuberculosis, can pose serious health threats to medical personnel. Creation of a "hard" structure may be necessary, therefore, to allow closed ventilation of these wards and to maintain conditions necessary to protect medical personnel and other patients from highly contagious and serious diseases.

Additional laboratory testing capabilities, however, are not necessarily required for these operations. For example, in the case of AIDS, positive test results would not necessarily affect a patient's treatment nor the precautions the medical staff would undertake. However, given the high rate of HIV in some civilian populations and among some coalition forces, it will be critical to address the concerns of the medical staff as to the risk of exposure in these operations.[15] The AMEDD, therefore, will want to continue to be proactive in terms of education, advance preparation, and counseling of deploying personnel in addressing the risks of exposure to certain serious infectious diseases. In addition, nonmedical personnel will need to be educated on how to minimize their risk of exposure to such diseases.

Injuries

Because some of the most common types of injuries in OOTW are sports-related, the Army may want to implement an aggressive preventive medicine and physical therapy program for these operations. Such a program could minimize the number of orthopedic injuries, increase the rate of return to duty, and reduce the demand for orthopedic surgery and outpatient physical therapy services in-theater.

As land mines proliferate and pose an increasingly serious medical threat, the AMEDD can expect a continued need to treat complex mine injuries in a field setting. How this may alter the support requirements for OOTW needs to be evaluated. For example, it could be that additional traction capabilities, limited rehabilitative services, and extended physical therapy capabilities in-theater may be required. Particularly in coalition operations where repatriation problems may remain intractable, the AMEDD can expect to continue to face the dilemma of managing the care of these soldiers in-

---

[15]The high rate of AIDS hypothesized among the Haitian civilian population (60 percent) raised a number of serious concerns among U.S. military medical personnel. Despite U.S. medical policy, there was a significant amount of treatment of Haitian civilians that was unavoidable. The fear among the medical personnel was real, and much counseling was needed to explain why treatment of some civilians was necessary, why it was part of their mission, and what precautions could be undertaken. As summarized by the former commander of the 55th Medical Group, the problem was that treating a civilian with AIDS from the medical staff's perspective meant not just risking infection to one's self, but also risking exposure of his or her family to the AIDS virus.

theater longer than what would normally be expected; this needs to be planned for.

Ensuring a Surge Capacity

Clearly, the size of the military hospital needed for OOTW tends to be small. This means that an Army hospital may easily be overwhelmed in a mass-casualty situation. Not only do planners need to take this fact into consideration, but in addition the Army and the AMEDD may want to consider developing a staging team along the lines of the Air Force's MASF model that would be capable of receiving casualties for initial treatment and then evacuating them out. Such a team could be co-located with the hospital or centrally located and capable of rapid response in the event of mass casualties. Staging teams would formalize the ability of Army medical units to deal with a surge in the event of a near-overwhelming or overwhelming casualty situation and serve as a means of redistributing patients quickly and safely.

Medical Logistics

As medical units are increasingly tailored to meet varied support requirements, medical logistics will face a difficult set of challenges. Tailoring a military hospital and other deployed medical units for OOTW expands the medical support requirements and complicates the logistician's job, since stock items may no longer meet the needs of a specific operation. As a result, these operations tend to be more resource intensive, personnel intensive, and difficult to predict in terms of requirements for medical supplies and equipment. Given the broad range of operations the Army will be called upon to support in the future, more flexible short- and long-term planning strategies will be needed by the medical logistics community.

Supporting a Broader Patient Population

The AMEDD has created support packages, including sets for pediatric and geriatric patient populations, for use in OOTW. For obstetrics and gynecological care, the AMEDD needs to incorporate similar support packages. To date, however, the care the AMEDD has provided these three patient populations has tended to be underestimated. Therefore, it will be critical that planners, medical commanders, and their staff become aware of the need to incorporate such packages as part of the support requirements and better anticipate the nature of the patient population to be served.

If the United States defined its strategic medical objectives more broadly to include some provision of civilian or refugee care, Army hospitals would face additional requirements. At a minimum, an Army hospital would need to be able to set up separate wards for pediatrics and civilian adult patients (i.e., separate wards from those housing coalition forces). It also would require a pediatrician or family practitioner and a limited range of pediatric equipment and medical supplies. Pediatric patients also require housing for adults who accompany a child. In several recent instances, U.S. military hospitals have ended up housing orphan children. The hospitals would need guidelines for determining the type and range of care they should provide such patients in a field setting. Finally, they would probably need a medical officer who could serve as a liaison with UNHCR, the local medical community, and the local government to coordinate the provision and transfer of these patients' care.

Telemedicine

Telemedicine has the potential to play an important role in OOTW, where many different specialties may be required to treat a wide range of diseases and medical conditions—all of which cannot be covered by any single medical element. Because OOTW tend to have a ceiling imposed on the total number of U.S. troops, after factoring in the various components of a force, the medical component often is highly constrained in the number of medical personnel that may be deployed. As a result, some have proposed that telemedicine may offer a means to reduce the size of the medical requirement in-theater. However, telemedicine is not necessarily likely to save any "in-country spaces," since even if a few provider spaces are saved, the technical people needed to run the system may cancel out any savings on the provider side. Further, as illustrated by Somalia, a certain minimum level of medical support is required in-theater in order to ensure a surge capability.[16]

There are also technological and operational issues that remain to be addressed in evaluating telemedicine's potential. For example, in the event of a "hot spot" or a combat scenario when the rest of the Army or the JTF is burning up the satellite links with C4I needs, whether medical units will be able to get the bandwidths they need to do quality telemedicine work is an important question to be addressed.

In addition, there are some innovative uses of this technology that remain to be explored. For example, telemedicine may be able to play a significant role in addressing repatriation problems, as well as in providing medical intelligence and linguistic requirements. For instance, a direct link to the embassies of those countries who have contributed troops for a multinational force may help facilitate the evacuation of coalition patients. In terms of medical intelligence requirements, such a capability would allow U.S. military physicians to obtain guidance on ethical issues (e.g., do-not-resuscitate orders on a soldier who has incurred a serious brain injury) or treatment decisions from a soldier's own military medical department. In terms of linguistic requirements, one could envision a military hospital having the capability to talk with language experts within CONUS or to a foreign military physician or nurse stationed in a soldier's source country to facilitate treatment decisions.

However, to date the use of telemedicine capabilities in the theater has been limited primarily to the transmission of images back to fixed facilities within CONUS or Europe and to teleconferencing. Certainly, the full range of this technology's potential has not yet been realized. As the Army and other services move forward with adopting this technology, it will be important to understand better both its potential and its limitations on the battlefield.

---

[16]Interview with COL Carroll, Army War College, November 1995.

MINIMIZING THE IMPACT OF OOTW ON AMEDD READINESS AND ON PEACETIME CARE

Continued support of OOTW has the potential to stress the Army health service support system, affecting both future wartime readiness and peacetime health care delivery to beneficiaries. As discussed in Chapter Five, it is not that any single OOTW is demanding in terms of large numbers of medical personnel or units required. Rather, what makes OOTW challenging is the simultaneity of demands, the fact that these operations tend to be open-ended, and the Army's direction that it support these operations without any degradation in beneficiary care. For example, to meet past OOTW demands, the Army has deployed key elements of some hospitals and has pulled individuals from several other military treatment facilities (MTFs) in order to do so. This in turn affects the entire peacetime health care delivery system.

This might seem to suggest more use of the Reserve Components, where much of the medical structure is located. However, it is difficult to augment deploying units with reservists or to backfill hospitals with individuals in the right specialties when unplanned OOTW missions arise. There are also a number of constraints in the employment of reserve medical assets in these operations, suggesting that the active component likely will continue to be responsible for the bulk of the medical support in future OOTW.[17]

To preserve its capabilities in the face of OOTW demands, the Army may want to consider designating certain medical units as OOTW hospitals and staffing those hospitals with two of each of the most critical functional elements. For example, of the 13 CONUS TOE hospitals currently in the active-duty structure, the AMEDD could build one or several into a "1.5" hospital. Then, if half of the hospital deploys on an OOTW, a complete hospital will still be available for a second deployment.[18] Such units would then know in advance (for a

---

[17]L. M. Davis, G. Hepler, and R. A. Brown, *Assessing the Use of Reserve Medical Forces in Operations Other Than War,* Santa Monica, CA: RAND, MR-817-OSD, 1996.

[18]Although up until now there has been little intent by the AMEDD to split a unit apart and to have the two pieces capable of undertaking independent missions—i.e., operate simultaneously in two different places, the AMEDD may want to reconsider this policy in terms of OOTW. Given the reduction in the number of active-duty

one-year period, for example) that they would be on the "hot" seat for supporting OOTW. This designation could be rotated among existing AC hospitals on a yearly basis.

Such rotating designations would provide an element of predictability, but also open up the possibility for real advance planning for these missions. Under this proposal, the Army could avoid pulling personnel from a number of different MTFs to support a single deployment and, thus, degrading services across the entire peacetime health care delivery system. Also, this proposal would enable individual Army MEDCENs and MEDDACs (from which PROFIS personnel are pulled) to do advance planning to maintain beneficiary care while supporting a deployment. For instance, they may choose to negotiate standing contracts with civilian providers or place deployable PROFIS personnel in noncritical positions to minimize the impact on peacetime health care when they are deployed.

COALITION OPERATIONS

As discussed in Chapter Five, whenever the United States is involved in UN operations or with a multinational force it will encounter some unique problems in terms of providing and structuring the medical support for these operations. Instead of being able to set up an integrated structure of echelons of care with consistent quality, the U.S. military will potentially face a hodgepodge structure with holes and gaps and of variable quality.

Clearly, the U.S. military tends to serve as the backbone of the medical support in multinational operations. Partly this has been because the United States has the best (and the most expensive) medical support available. As a result, historically our allies often have relied on us for medical support, whether it be an explicit or implicit part of the mission. However, in OOTW it is also clear that the United States is the driving force behind much of this in that we impose our own standards on other forces and drag the UN and our coalition partners along with us. Given this, it is up to the United States to put forth a

---

hospitals in the overall force structure, the increasing number of OOTW to support, and the problems encountered in utilizing reserve medical units for these missions, Army medical support for OOTW in the future may necessitate greater flexibility as recommended here.

set of solutions that it can live with to define its medical policy in coalition operations. Ideally the United States would secure an agreement with other nations and seek to promulgate the plan through the UN or other multinational organizations.

Echelons of Care

The United States and its other key coalition partners (e.g., Britain and France) may want to take the lead in developing a revised definition of echelons of care, specific to OOTW involving a multinational force. The traditional operational (or "wartime") definition of echelons of care has not worked well in recent OOTW. In wartime medicine, the objectives are rapid intervention, life sustainment support, and evacuation back or airlift out of the theater to a more definitive level of care; in contrast, in peacetime care a physician is able to bring to bear a full range of expertise, medical supplies, equipment, and support personnel to provide comprehensive care to a patient. In UNPROFOR, the British and French utilized the theater medical system appropriately as in wartime, quickly evacuating their injured and sick soldiers out of the theater. If the United States had had troops on the ground or had not been the main provider of health care for UNPROFOR, it would have done the same. The developing countries, however, did not use the theater medical system as it was intended and, instead, utilized U.S. military hospitals more like community hospitals in a peacetime setting. Coupled with repatriation problems, this led the United States at times to provide Echelon IV care for the UNPROFOR force and to treat a disproportionately greater number of troops from developing countries than from other nations.

A draft UN plan or concept for medical support in multinational operations needs to be developed. Such a plan would set standards in terms of medical readiness, unit readiness, training, equipment, and standards of care. It would also need to address such issues as: Should the principle be equal access to the same level and quality of medical care for all forces in these operations? If so, then how can one accomplish this without favoring one set of troops over another and without getting into the provision of peacetime health care in a theater of operations?

One option is that the U.S. military and its coalition partners develop alternative definitions of echelons of care for OOTW. For example, Level I could be defined as providing treatment to military forces only and evacuating them as soon as possible. Level II could treat military forces only for up to three weeks (including minor surgery and emergency care) and then evacuate them. And Level III could include hospitalization for military and civilian patients, including some rehabilitative services, to be provided by civilian contractors. This option would be undertaken primarily for political reasons (e.g., where we elected to treat civilians or decided that we could not accept having two different standards of care for coalition forces in OOTW).

Alternatively, the United States and its coalition partners could set up a policy on echelons of care that says to the UN that Echelon II is as far as we are willing to go and that for other care we expect the UN or the coalition itself to establish contracts with fixed facilities in neighboring countries. Under such an arrangement, soldiers whose own countries lack adequate evacuation resources or are unwilling to repatriate their injured can be transported to these facilities for more definitive care, instead of remaining in-theater. Without such an arrangement, we could face a two-tiered system of care, one for Western forces and another for troops from developing countries, which would probably not be politically sustainable.

The United States has not needed a repatriation policy in the past. However, in the case of OOTW involving a multinational force, it may need to incorporate one as part of the formal mission statement in future operations.

If echelons of care are not redefined, an alternative option may be for the U.S. military to serve as the coordinator of medical care in these operations. In this way, we could ensure that the quality of theater medical assets and the functioning of the health service support system was maintained. This is a limited solution, though, since it does not address the inadequacies in other coalition forces' medical assets, variations in quality of those assets, and the ill-preparedness of some troops. Or the United States may want to continue to impose its standards on other coalition forces in terms of echelons of care, equipment and supplies, training, standards of care, and medical readiness. Doing so could entail training and equipping other forces' medical assets for a specific operation. If we choose either of the above options, then we need to be explicit about it and negotiate compensation up front from our coalition partners or the UN.

Standards of Care

Differences in standards of care and medical practice from country to country pose related questions: Can the United States avoid plugging the holes in the theater medical system? If so, how much variability in the theater medical assets can we afford? What risks do we run by doing so? For example, some militaries have lower standards of care than the United States, particularly in such areas as trauma care, where the United States tends to be far more aggressive.[19] Although we may be able to maintain quality control in a clinic setting by teaming up U.S. military physicians with the medical staff of forces from developing nations (as was done during the MFO Sinai peacekeeping mission), we may not be able to do so in an emergency situation—where the first assets to reach a wounded U.S. soldier may be a medical team from a poor country. In such cases, the standard of care delivered may not match normal U.S. expectations. Does the risk of such incidents mean that the United States cannot afford to allow much variability in theater medical assets in multinational operations whenever U.S. troops are on the ground? If so, then how can coalition medical assets be integrated into a theater medical system such that U.S. troops are protected and the same high quality of medical care is provided to the entire multinational force?

Individual and Unit Readiness of Coalition Forces

When the U.S. military has deployed for OOTW as part of a multinational force, it has often failed to realize the shape other countries' forces were in. Some coalition partners proved to have low levels of individual medical readiness and unit readiness. An overall U.S. policy for dealing with the UN and coalition forces must deal with these readiness problems. In the past, we have reacted by plugging

---

[19]Interview with COL Smerz, USSOCOM Surgeon; Health Care Operations Conference, San Antonio, TX, June 1995.

the holes in the theater medical system caused by ill-prepared units from other countries—often without the possibility of reimbursement. Further, U.S. military medical units then faced the problem of not necessarily having the right mix of specialists or the right amount or configuration of medical units, equipment, or medical supplies.

In addition, because U.S. military hospitals end up treating disproportionately more troops from developing countries than from other nations, U.S. military medical personnel run a higher risk of exposure to serious infectious diseases (some of which cannot be immunized against) than other troops. This raises several questions: How can the United States ensure the safety and health of U.S. military medical personnel and troops participating in OOTW? For this reason alone, should we only provide medical care to U.S. troops in these types of operations? If this is not politically feasible, does the United States need to insist on standards of medical readiness for all troops comprising a multinational force?

Air Evacuation and Logistics

Regardless of the policy or formal arrangements, the U.S. military can expect to be tasked increasingly to provide MEDEVAC and medical logistics assets in UN-led or informal coalition operations. This stems from the fact that the United States has one of the few militaries with these capabilities. As illustrated by the experience of all three services during missions to the Balkans and Haiti, the military relied on U.S. support over UN systems because of quality problems and differences in standards. Given this, there needs to be a better mesh between logistics and medical units in these operations. Until the inadequacies of the UN medical logistics system can be addressed, U.S. medical units must continue to rely on U.S. supply sources in coalition operations, regardless of the formal tasking. While some coordination may improve this situation, we expect a continued demand and reliance on these U.S. assets, which should be planned for.

Maintaining the Blood Supply

Maintaining the blood supply will continue to be an important concern in OOTW undertaken by a multinational force. Although each coalition partner during UNPROFOR was to be responsible for its own blood supply, in reality only the Western countries were capable of doing so. U.S. policy has been to not use other countries' blood, even when treating coalition soldiers, due to the fact that some countries do not routinely screen for some HIV-related viruses.

In addition, one's ability to tap into the civilian blood supply may be limited in these operations and dependent on whether the local populace itself has a high demand for blood (e.g., because of a larger number of civilian casualties). As seen during UNPROFOR, land mine injuries alone may quickly use up a military hospital's blood supply. This problem, along with concerns about the quality of other countries' screening procedures and cultural sensitivities about who is receiving whose blood, led the CINC to implement a frozen blood program during UNPROFOR. Routine inclusion of such a program in future operations may be necessary.

Security

Security of a military hospital and of its medical staff is an important concern in OOTW, especially those operations involving coalition forces or UN missions. In some instances, a U.S. hospital may be the sole U.S. presence in the theater and thus responsible for all of its force protection needs. In other instances, it may rely on the UN or coalition troops for some force protection. Further, in some OOTW there may not be a "rear" where the hospital can be located.

In addition, because U.S. military personnel are a high-visibility target, it is critical to provide for the security of the hospital and individual medical personnel who may undertake sector visits, MEDEVAC missions, or outreach programs within the local community or to other coalition forces. Security concerns led to tight restrictions on the movement of U.S. military medical personnel within the theater during recent OOTW. Further, as was the case during UNPROFOR, the UN may not always be as responsive to U.S. force protection concerns as one might expect. Other coalition troops also may not provide the level of force protection considered necessary by U.S. standards; for example, non-U.S. forces were responsible for the security of the hospital compound's perimeter in Mogadishu, but there were concerns about the reliability of those troops.

In general, U.S. medical units should be prepared to provide security for a hospital compound's perimeter and take care of their own force protection needs in OOTW. This has implications in terms of the training requirements for OOTW, as well as staffing implications, since a certain percentage of the medical personnel may be tied up with security functions rather than medical functions at any one time.

Providing Training and Support to Local Health Care Providers

If the United States defines its strategic medical objectives more broadly to include its military working with the local community in reestablishing or improving the medical system, or assisting relief agencies in becoming self-sustaining, the U.S. Army would need to bring in additional medical equipment and supplies. For example, even though the United States may not be called upon to provide direct medical care to civilian populations, the U.S. Army may be asked to supply a generator or other medical equipment to help a local health clinic become operational again.[20] During UNPROFOR, for example, preventive medicine officers helped Sarajevo to ensure the quality of its water supply and thereby helped to avert an outbreak of cholera in the city.

Other activities may involve training local medical personnel. For example, in Haiti, U.S. military hospital staff did some training of their Haitian civilian counterparts in the local community hospitals. If such activities are to be supported, training materials, engineers, preventive medicine teams, and community health nurses may be required in future operations. If the Army included as part of its medical mission educating other coalition troops on basic preventive medicine and public health measures, in order to minimize these troops' demand for health services in-theater, then community health nurses and public health officers also would be needed.

---

[20]Providing equipment and supplies, however, runs the risk of these items being sold on the black market. This occurred in several recent operations. The AMEDD instead may want to restrict provision of such materials only to relief agencies and to work with the local hospitals primarily in a training capacity and in the coordination of care of civilians.

UN Accounting and Reporting Requirements

Finally, the UN's unwieldy bureaucracy and reporting systems have presented significant problems for the theater commander, JTF surgeon, and their staff in the past. To alleviate such problems, the Army may want to have comptroller support during the initial phase of a UN deployment. Such support would be responsible for figuring out the UN system of reporting and accounting and for establishing a viable system for the combined or joint task force. The comptroller would not necessarily need to be in-theater for the duration of the rotation, but long enough to help get the system up and running efficiently. Class A agents and core staff could then be trained on that system.

OVERALL OBSERVATIONS

As the AMEDD, like the rest of the U.S. military, continues to downsize, no one can clearly envision the strategic environment for the future. We do anticipate, however, that the United States will continue to undertake OOTW, perhaps at an increasing rate.

In this report, we examined how the AMEDD may ensure broadbased flexibility to support the diversity of new missions it faces in OOTW and coalition environments. Most of the issues identified are not unique to UN operations, but also will apply to other multinational operations, such as the current NATO peacekeeping mission in Bosnia.

In general, peacetime OOTW entail a broader set of demands upon the medical component. Planning for future OOTW needs to recognize the breadth of such demands and not assume that they will be limited to the traditional support requirements of combat forces. The medical issues associated with coalition operations, in particular, are complex and have implications for the overall success of the military mission. As seen in our two case studies and in other recent operations, the United States needs to focus and contain its medical involvement in these missions where possible. Finally, many of the medical issues identified here are systemic—to be confronted successfully, they need to be addressed not only at the AMEDD headquarters level, but also at the strategic, operational, and tactical levels.

DATA AND METHODS FOR PATIENT AND PROVIDER-LEVEL ANALYSES: UNPROFOR AND SOMALIA

UNPROFOR DATA AND METHODS

For the patient utilization analyses presented in Chapter Three, we drew from several different data sources. For the two Army Mobile Army Surgical Hospitals (212th MASH and 502nd MASH), we were able to obtain patient-level data from the Directorate of Patient Administration Systems and Biostatistics Activities (PASBA), AMEDD Center and School. These data covered the period between November 1992 and October 1993. Data elements included number of outpatient visits, number of admissions, and length of stay for each patient category.

Since our interest was also in comparing differences in utilization between U.S. personnel, foreign military, foreign civilians, and UN/NATO employees and officers, for this analysis we further grouped the Army patient data into these four categories. Assumptions made in determining these groupings were similar to those described below for the Somalia deployment. The only difference in patient groupings between Somalia versus UNPROFOR and Provide Promise is that for the latter deployment, UN personnel and NATO employees and officers were combined into a single category.

To examine overall differences in utilization across the four rotations, we present patient-level data obtained from briefing charts put together by the Navy's Fleet Hospital 6, which summarized for each rotation the number of outpatient visits, number of admissions, and proportion of inpatients with disease versus trauma-related conditions. For this comparison we needed information on both the Air

135

Force and Navy hospitals that undertook the third and fourth rotations into Zagreb, in addition to those rotations done by the U.S. Army. This information could only be obtained from briefing charts. In addition, we utilized unpublished length of stay data for the Navy's Fleet Hospital 6 obtained from CAPT James Carlisle, Chief of Clinical Services, in order to compare how average length of stay differed across contingents during the fourth rotation. We used data for the Navy hospital for this comparison because the PASBA data for the two Army hospitals did not allow us to break down the foreign military category by individual contingent.

SOMALIA DATA AND METHODS

Provider-Level Analysis

To track medical support for the mission in Somalia, we use data on Joint Task Force–Somalia (JTF-S) Professional Fillers (PROFIS) personnel. These data do not include the organic medical assets belonging to the 10th Mountain Division (the division that served as the backbone of U.S. forces in Somalia). The organic medical assets of a field unit or division typically include physician assistants (PAs), medics, and Medical Service Corps (MSC) officers who are regularly assigned to a medical field unit on a full-time basis.[1] Physicians, nurses, and other specialties will mostly be designated as PROFIS personnel, with these individuals spending most of their time, when not deployed, in a fixed facility. Therefore, although the PROFIS personnel included in this analysis represent the bulk of the medical support in Somalia, they do not represent all of it, since a few individuals, physicians for example, were assigned full time to the 10th Mountain Division as part of the division's organic medical assets.

Since AMEDD personnel came and went at various times during the Somalia deployment, we counted the number of PROFIS personnel at the midpoint of each month. This allowed us to obtain a consistent snapshot of what the medical support looked like for each month of the deployment and how it changed over time.

---

[1]Physician assistants (PAs) are assigned to a medical unit attached to a division full time, in theory. In actuality, because they also need to see patients, they will spend some of their time in a fixed facility.

We began with data on the total number of PROFIS personnel deployed to Somalia by unit and area of concentration (AOC) occupation. The number of PROFIS personnel at the midpoint of each month is shown over the course of the entire operation, starting with November 1992 through March 1994. Certain AOCs were grouped as follows: under the Preventive Medicine Officer category, preventive medicine physicians, entomologists, and environmental science officers were grouped. Administrative, logisticians, and operations officers were included under the health services officer category (i.e., those officers involved in administration or the operational aspects of a medical field unit).[2] The behavioral sciences category included psychologists and social workers.

For this analysis, because we wanted to examine how the specialty mix changed over time, we further grouped the AOCs into the following specialty categories: preventive medicine, primary care and medicine, surgical and related specialties, mental health, dental, nursing, administration and health services, and other specialties. Table A.1 lists the specialty categories and the AOCs that fall within each grouping.

Interpretation of the AOCs or specialty mix has to be done with caution. While a PROFIS individual is designated to fill a specific slot in a deploying medical unit, his or her MTF commander has a fair amount of leeway in terms of who actually may be deployed. If a commander cannot afford to lose a particular individual, he may send instead another to fill the PROFIS slot. In addition, the AMEDD has recently revised some of its AOC codes, which has made the interpretation of some of the PROFIS taskings ambiguous.

Patient-Level Analysis

To examine changes in patient utilization over the course of the Somalia mission, we obtained patient-level data from the Patient Administration Systems and Biostatistics Activities, MEDCOM. The

---

[2]There are few PROFIS individuals in health services officer slots, since usually these AOCs will be organic to the division and the medical unit. Therefore, it is less common to augment these administrative-type positions via PROFIS.

Table A.1

Specialty Categories and AOCs—Somalia

Specialty	AOC
Preventive Medicine	
Preventive Medicine Officer	60C
Community Health Nurse	66B
Preventive Medicine Officer	67C
Primary Care and Medicine	
Pediatrician/Internist	60P,61F
Family Practitioner/ER/PA	61H,62A,65D
Flight/Field Surgeon	61N,62B
Surgical and Related	
Ob/Gyn	60J
General/Thoracic/Orthopedic	61J,61K,61M
Other Surgical Specialties	60N,60S,60T,60Z,66F
Mental Health	
Psychiatrist	60W
Mental Health Nurse	66C
Behavioral Sciences Officer	67D
Dental	
General Dentist	63A
Comprehensive Dentist	63B
Oral Surgeon	63N
Nursing	
Operating Room RN	66E
Medical-Surgical/Clinical RN	66H,66J
Administrative and Health Services	
Executive Medicine Officer	60A
Medical Maintenance Officer	670A
Health Services Officer	67A
Aeromedical Evac Officer	67J
Other	
Pulmonologist	60F
Infectious Disease Officer	61G
Diagnostic Radiologist	61R
Dietitian	65C
Laboratory Sciences Officer	67B
Pharmacy Officer	67E
Optometrist	67F

in-theater data had been collected by the individual hospital units and then reported to the MEDCOM.

The patient data cover the period between January 1993 and January 1994. Note that the data do not cover the initial few months (November and December 1993) or the latter few months (February and March 1994) of this operation. This is in contrast to the provider data, which covered all 17 months of this deployment. Also note that patient data were unavailable for April 1993 and August 1993, the two months when the rotation of U.S. troops and hospital units into the theater took place.

We were able to obtain data on outpatient visits, admissions, length of stay, clinic of service (or disposition), and patient category. Since our interest was in comparing differences in utilization between U.S. personnel, foreign military, and foreign civilians, for this analysis we further grouped patients into these three categories, as shown in Table A.2.

In terms of the foreign civilians and other foreign nationals, the three hospital units were not consistent in how they coded patients between these two categories, so we combined them into a single category. The foreign civilian category we use consisted mostly of Somali nationals, with a few non-U.S. civilians (e.g., relief workers).

We were also interested in comparing the distribution of patients across clinical services within a hospital. Table A.3 shows the four groupings we used to examine the inpatient services and lists the type of visits categorized under the outpatient listing. We were unable to separate out emergency room visits from other types of outpatient visits, since the coding was inconsistent across the three types of hospital units. For a separate analysis we also compared the distribution of surgical patients across the different surgical specialties, using data on patients' disposition (i.e., the last clinic of service) (see Table A.3).

Table A.2

Patient and Clinical Services Categories—Somalia

U.S. Personnel
   Military
        Army active duty
        Navy active duty
        Marine active duty
        Air Force active duty
   Civilian
        Federal department employee
        Federal agency employee
        Dependent, non-DoD federal agency
        Contract employee

Foreign military
   Foreign military
   NATO military personnel
   Non-NATO military personnel

Foreign civilians
   Foreign civilian
   Other foreign nationals

Table A.3

Clinical Services and Surgical Disposition Categories

Clinical Services	Surgical Disposition
Inpatient admissions	General/thoracic
Internal medicine	
Internal medicine	Orthopedic
Infectious diseases	
Family practice medicine	Other surgery
	Obstetrics
Surgery	Vascular
General	Otorhinolaryngology
Orthopedic	Neurosurgery
Neurosurgery	Oral
	Head and neck
Ob/gyn	
Obstetrics/gynecology	
Family practice obstetrics	
Psychiatry	
Psychiatry	
Family practice psychiatry	
Outpatient visits	
Emergency room	
Orthopedic outpatient/casts	
Mental health/social worker	
Primary care/family practice	
Acute minor illness visit	

SOMALIA DEPLOYMENT DATA

Interpretation of the AOCs or specialty mix for the Somalia deployment needs to be done with caution. A PROFIS individual may be assigned to fill a designated slot in a medical unit, but his or her MTF commander has a fair amount of leeway in terms of who actually will be deployed. The AOC where this is most likely to occur is the position of Field Surgeon (62B). There is no medical specialty in the peacetime structure that corresponds to this position. Rather, a 62B is a combat or field position, and a variety of different medical specialties may fill this slot. For example, in Table B.1 the Field Surgeon slot may represent an internist, a pediatrician, a general surgeon, or in rare instances, a resident.[1]

In addition, the AMEDD has recently revised some of its AOC codes, which has made interpretation of PROFIS taskings somewhat ambiguous (Table B.1). The problems concern the MSCs and PAs who were once warrant officers in the Medical Corps, but have now been reorganized as commissioned officers in the Medical Specialist Corps. In both cases, a mixture of the old and new AOC codes were named as requirements for PROFIS fillers.

The reorganization of the Medical Service Corps' AOCs from a three-character alphanumeric code to a five-character alphanumeric code presented the most challenging problem, for some of the old 68-series AOCs were changed to codes that were indistinguishable from the old 67-series AOCs. Use of the old AOC codes resulted in some

---

[1]The AMEDD tries not to touch GME for deployments. In Desert Storm/Shield they did have to tap residents, but this was an unusual instance.

confusing results, such as the deployment of Health Services Information Managers (old 67D) rather than Social Work Officers and Psychologists (new 67D). Similar concerns involved comptrollers and preventive medicine officers, field medical pharmacists, and personnel managers and optometrists. Whenever possible, we interpreted the AOCs according to the new codes, which has the potential of overrepresenting laboratory science, preventive medicine, behavioral science, pharmacy, and optometry officers, while underrepresenting taskings of health services administrators, comptrollers, information managers, patient administration, personnel, and plans and operations officers. This procedure seemed to be the most reliable, since the old 67-series Medical Services Corps positions, now recoded as 67A or 70*67,[2] are usually filled by officers actually assigned to these units, rather than by PROFIS fillers. The old 68-series positions, now coded as 67B, C, D, E, F, and G (among other more definitive codes), are usually filled by officers who spend most of their time working in hospitals and other medical activities, but might be designated as PROFIS fillers.

Table B.1 shows a comparison of the old and new AOCs for the Medical Service Corps. The new AOCs 67A through 67D are catch-all AOCs and include the group of definitive AOCs listed below each category. Assumptions we made in categorizing individuals under the old and new AOCs were as follows: Health Services System Management Officers (67D), computer support personnel, were deployed to support the telemedicine capability. Patient Administration Officers (old 67E) and Health Services Human Resource Managers (old 67F) are typically organic to a medical field unit, so it would be unusual to have PROFIS personnel assigned to these slots. Therefore, we assumed these codes represented the new AOCs of Pharmacy Officer (67E) and Optometrist (67F).

---

[2]70*67 refers to the following new AOC series: 70K67, 70A67, 70B67, 70C67, 70D67, 70E67, 70F67, 70H67.

Table B.1

Revisions to the Medical Service Corps'
Areas of Concentration (AOCs) Codes

Medical Functional Area (MFA)	"Old" AOCs	"New" AOCs
Health Services Officer (IM)	N/A	67A00
Health Care Administration	67A	70A67
Health Services Admin Asst	67B	70B67
Health Services Comptroller	67C/67G	70C67
Health Svc Sys Mgt Officer	67D	70D67
Patient Administration Officer	67E	70E67
Health Svc Human Resources Mgr	67F	70F67
Health Svc Plans, Intel & Training	67H	70H67
Health Services Material Officer	67K	70K67
Aeromed Evacuation Officer	67J	67J00
Laboratory Sciences Officer (IM)		67B00
Microbiologist	68A	71A67
Biochemist	68C/68J	71B67
Parasitologist	68D	71C67
Immunologist	68E	71D67
Clinical Laboratory Officer	68F	71E67
Health Svc Research Psychologist	68T	71F67
Preventive Medicine Officer (IM)		67C00
Nuclear Medicine Science Officer	68B	72A67
Entomologist	68G	72B67
Environmental Science Officer	68N	72D67
Sanitary Engineer	68P	72E67
Audiologist	68M	72C67
Behavioral Sciences Officer (IM)		67D00
Social Work Officer	68R	73A67
Clinical Psychologist	68S	73B67
Pharmacy Officer	68H	67E00
Optometrist	68K	67F00
Podiatrist	68L	67G00

Note: 60C Preventive Medicine Officer is a physician; 67C Preventive Medicine Officer is a nonphysician and includes audiologists, sanitary engineers, environmental science officers, entomologists, and nuclear medicine science (radiation protection) officers.

EDUCATION AND TRAINING INITIATIVES

There are a number of initiatives under way to incorporate OOTW into the education of AMEDD officers and enlisted personnel. Currently, AMEDD active-duty personnel receive some education on OOTW, although not all personnel get exposure nor receive the same level of detail, depending on where an individual is in his or her career and on the specific course.

Junior and senior enlisted personnel do not receive any instruction on OOTW in either the Basic Non-Commissioned Officer Course (BNCOC) or the Advanced Non-Commissioned Officer Course (ANCOC) or at the Sergeant Majors Academy.[1]

AMEDD officers in the Officer Basic Course (OBC) receive a three-hour block of instruction on OOTW, with one-third of this time devoted to OOTW fundamentals (emphasis on environment, principles, and activities of the Army in operations other than war) and the remaining two-thirds focused on domestic support operations (the National Disaster Medical System (NDMS) and management of emergency health care during a national disaster).[2]

At around the 3- to 5-year mark in an officer's career (senior 1st lieutenant, junior captain), he or she attends either the AMEDD Officers' Advanced Course (OAC) or the Combined Logistics Officers'

---

[1]Interview with Ms. Jackson, Instructional Systems Specialist, AMEDD Non-Commissioned Officer Academy, 8 December 1994.

[2]Interview with CPT Judith Robinson, Medical Operations Instructor, Medical Operations Branch, AMEDD Center and School, 9 December 1994.

Advanced Course (CLOAC). In the OAC (26-week course), officers receive 31 hours of instruction on OOTW, including two hours on low-intensity conflict as described in FM 100-5, three hours on CONUS OOTW operations, three hours on OCONUS OOTW operations, six hours of student briefings, and seventeen hours of OOTW practical exercise.[3] As this course is currently being reengineered and shortened to 10 weeks, the AMEDD should ensure that this part of the curriculum remains strong. CLOAC also includes classroom instruction and a practical exercise on medical support for the full range of military operations, including OOTW.

The Combined Arms and Services Staff School (CAS3) comes at the 8- to 10-year mark in an officer's career and is aimed at preparing individuals for staff officer positions. CAS3 is taught in two phases, with Phase 1 being a 140-credit-hour correspondence course and Phase 2 taught at Fort Leavenworth, Kansas over the course of nine weeks. The only OOTW-type instruction presented during CAS3 is a two-day contingency operation planning exercise during the residence phase.

In the Command and General Staff Officer Course (CGSOC) (at around the 11- to 13-year mark in an officer's career), officers are trained to become field grade commanders and staff officers principally at the division and corps levels. For the active component, this course is taught in three terms, with 36 contact hours during the second core phase being devoted to OOTW operations. Subject matter includes an overview of the environment, root and cause of conflict, senior-level leadership in the joint arena, training for OOTW, introduction to OOTW analysis, counterinsurgency operations, security assistance, humanitarian and disaster relief operations, and peace support operations.[4] During the second and third core phases, two OOTW elective courses are also available on health service support in force project operations (27-hour course emphasizing the joint medical support in OOTWs)[5] and logistics in operations other than

---

[3]Ibid.

[4]Interviews with LTC Wyssling, 14 December 1994; LTC White, Instructor, LTC Swan, Canadian Exchange Officer, and Mr. Babb, Instructor, Command and General Staff School, 12 December 1994.

[5]Interview with LTC Mokri, Instructor, Command and General Staff School, 14 December 1994.

war (emphasis on logistics issues unique to OOTW and UN operations).[6]

The Army War College is a year-long course designed to prepare senior officers for top leadership positions in the Army. Officers either attend a year-long course in residence at Carlisle Barracks, PA or complete a correspondence course that includes two 2-week resident phases at Carlisle Barracks. Two 3-hour blocks of instruction on OOTWs are presented in each option. The resident course also offers an elective on theater OOTW with emphasis on joint operations.[7]

In addition, exportable training packages have been developed by the AMEDD Center and School to be sent to AMEDD commanders of deploying units participating in an OOTW.[8] These separate training packages are available for each major type of OOTW and include information specific to the type of mission, operational details on health services support, law of land warfare and establishment of health service support policy, UN policy, and NGO interface.

In terms of combat training centers, at the NTC, medical play is limited to Echelons I and II; OOTW is not included. At the JRTC, although a number of the exercises deal with OOTW scenarios, they tend not to cover the range of OOTW medical issues outlined in this report. Further, although the medical forces that rotate through the JRTC go from platoon level (Echelon I) up through the hospital units (Echelon III), historically only one or two rotations out of twelve will involve Echelon III units. Also, because of budgetary constraints there is some discussion of curtailing altogether Echelon III medical unit participation at the JRTC.

Some of the coursework outlined above tends to be primarily descriptive in nature. We would contend that it is critical for the coursework to be problem oriented, focusing on specific issues and their solutions and including such topics as repatriation problems,

---

[6]Interview with MAJ Dotson, Instructor, Command and General Staff School, 14 December 1994.

[7]Interview with COL Stovall, Director, Training and Force Readiness, U.S. Army War College, 20 December 1994.

[8]Interview with CPT Thacker, Chief, Training Operations Branch, Individual Training Division, AMEDD Center and School, 6 December 1994.

ethical and treatment dilemmas, identification of and dealing with potential sources of mission creep, and understanding the critical determinants of the medical support requirements in OOTW. Commanders also need to be informed about UN organizations and procedures associated with UN-led operations.

Officers who typically command joint task forces have not been trained in depth on issues unique to OOTW missions to date. The reasons for this are severalfold. Today's senior officers (O-6 and above) typically have spent their Army career training and preparing for a major regional contingency or large-scale conventional warfare with the Soviet Union. Thus, all of their training has been geared toward this end. Also, the typical Army officer may have had little exposure or experience in dealing with the UN, NGOs, coalition forces, or other government entities (skills necessary in OOTW). Further, the military education system does not directly address the myriad of activities that must be accomplished by the JTF commander in an OOTW environment. Knowledge of UN organizations and procedures is one example.

To remedy this, several actions could be undertaken. One, the entire officer education system might include an *integrated* doctrinal approach to managing assets in an OOTW environment. Currently, only certain officers are receiving portions of OOTW doctrine in their military careers. The aim of the instruction should be to provide officers at all levels with the essential tools needed to plan, undertake, and lead such operations. This instruction should include lessons learned from recent OOTWs, as well as practical exercises. Two, education of noncommissioned officers (NCOs) in OOTW is also essential. Because of the role NCOs play in medical units, they too need to be aware of the OOTW principles outlined here and of the critical differences between operations other than war and combat missions.